HOW THE WORLD WORKS
BIOLOGY

HOW THE WORLD WORKS
BIOLOGY

*From the science of the ancients
to modern genetics*

Anne Rooney

ARCTURUS

ARCTURUS

This edition published in 2017 by Arcturus Publishing Limited
26/27 Bickels Yard, 151–153 Bermondsey Street,
London SE1 3HA

ISBN: 978-1-78428-663-7
AD004960UK

Printed in Malaysia

CONTENTS

INTRODUCTION: LIVING LIFE 6

CHAPTER 1 ANIMAL, VEGETABLE, MINERAL 10
Order, order • Paleolithic zoology • Classifying life • The Great Chain of
Being • The origins of zoology • Class system • Expanding kingdoms
• The end of hierarchy

CHAPTER 2 BEASTLY MACHINES 44
Before our eyes • Picture this • From organism to mechanism • Blood and
heat • How to build a body • Moving inside

CHAPTER 3 WHAT ABOUT PLANTS? 68
Plants at a glance • Water, soil, air – plant nutrition • Getting chemical
• In and out • Growing every which way • More plants

CHAPTER 4 SMALLER THAN SMALL 84
Imaginary small things • Smallish • Cell theory • In sickness and in
health • Even smaller

CHAPTER 5 NEW LIFE FROM OLD 106
Fertile dust • Starting at the beginning • Preformed or growing?
• Eggs abound

CHAPTER 6 THE BEST IDEA EVER 126
In the beginning . . . • A changing view of change • The evidence is under your
feet • The Earth moves • Evolution – now with dinosaurs • The missing link

CHAPTER 7 PARENTS AND PROGENY 156
The monk and the peas • Looking at cells • Mendel rediscovered
• From chromosomes to genes • Nailed: DNA and heredity are inseparable
• DNA revealed • Evolution and genetics come together • Eat your ancestor

CHAPTER 8 WE'RE ALL IN THIS TOGETHER 176
All one • All for one – and we're the one • The beginnings of
biogeography • The moving Earth • Living together • Ecology comes of
age • From biosphere to noosphere • The living Earth • Forward in time

INDEX 204

PICTURE CREDITS 208

Introduction:
LIVING LIFE

'When I have observed nature she has always induced me to deem no statement about her incredible.'

Pliny the Elder (AD23–79)

Biology is the study of life, and the story of biology is the ultimate narrative of discovery. It tells how we have explored the myriad forms that life can take and begun to comprehend the complexity of life on Earth. But the story so far is only the beginning. The closer we look, the more is revealed, like an unfolding fractal pattern.

From supernatural to natural

People observed, used and depicted animals for thousands of years before they could be considered to be engaged in biology. Our story starts in earnest in Ancient Greece, where the spirit of scientific enquiry was kindled. The natural world was unhitched from the supernatural, and independent thinkers sought explanations for the phenomena they observed. Early proto-biologists catalogued plants and animals and thought about how living things

'Nothing of any real consequence happened in biology after Lucretius and Galen until the Renaissance.'

Ernst Mayr (1904–2005), evolutionary biologist, 1982

work, reproduce and interact. Thus the foundations of botany and zoology were laid nearly 2,500 years ago.

Go slow

But then development stalled. The foundations of biology as a science were soon overlaid with unscientific thinking. From the end of the Classical period until the 16th century there was very little progress in the life sciences. Religious dogma – Christian and Islamic – replaced scientific enquiry in the West. Neither the intellectual climate nor the structure of society was conducive to the kind of free-ranging exploration that had got the sciences off to such a good start in Greece. Even in the Arab world, which made considerable advances in other sciences from the 8th century, biology did not prosper.

In Europe, the veneration of Classical authorities meant the teachings of Greek and Roman thinkers went unchallenged for centuries. From the Middle Ages, early errors were incorporated into the Church's teachings, and became so thoroughly entrenched that they could not be shifted without a major shift in world view. That shift came in the Renaissance.

The discovery of the Americas demonstrated irrefutably that new knowledge could be discovered about which the Ancients had said nothing.

Setting the scene for science

From the late 15th century, the Renaissance brought renewed confidence in human intellect and potential. Humanism, critical thinking and investigation replaced blind acceptance of authority, dogma and superstition.

A dismal view that the world was on a downward spiral had prevailed during the Middle Ages, but this no longer appealed. Europeans were discovering new lands, filled with new creatures, plants and possibilities. The Protestant reformer Martin Luther (1483–1546) had challenged the authority of the Catholic Church and begun the Reformation. Scientists began to discover laws that govern natural physical processes. The world was not so mysterious or immutably fixed as it had seemed. The invention of movable type in Europe led to the development of printed books, which meant knowledge could be disseminated more quickly and more accurately. Universities expanded, and the first dissection theatres opened in them. The great scientific project of the Modern Age had begun.

Enlightened at last

As the Renaissance merged into the Enlightenment, a rigorous spirit of scientific enquiry prevailed. The 16th century brought discoveries that overturned the way people saw the world. Copernicus demonstrated that the Earth is not the centre of the universe, but orbits the Sun. Microscopes and telescopes showed unimagined new realms. It became clear that in the past

THE CASE OF THE VEGETABLE LAMB

The mythical 'vegetable lamb' was believed to grow in central Asia. In one version, the plant had many pods, each of which burst open to reveal a lamb. In another version, a single lamb grew on a long, bendy stalk. It could tip the stalk over to graze on surrounding vegetation, but died when it had eaten everything within reach.

In 1557, the Italian scientist Girolamo Cardano pointed out that the plant could not gain enough heat from the sun to sustain the lamb, especially during the early stages of its development. Then, in the 1600s, vegetable lamb specimens sent to the Royal Society were proven to be fakes or, at least, not lambs.

people had believed a lot of nonsense. It was time to start correcting errors that were the legacy of blindly repeating ancient assertions for more than 1,000 years.

At last, old beliefs were overturned if they did not stand up to scrutiny. The royal societies founded in Europe in the mid-1600s set out specifically to research scientific subjects and determine the truth. The British Royal Society's motto, *Nullius in verba*, is taken to mean 'take nobody's word for it' and embodies the new, bold attitude.

SCIENTIFIC METHOD

Aristotle had encouraged his students to trust the evidence of the physical world, to enquire, observe, experiment and think critically – but this lesson was lost over time. Ironically, blind adherence to Aristotle's writings meant that people did not observe, investigate and think afresh as he had advocated.

In the 12th century, the English natural philosopher Robert Grosseteste was among the first Europeans to understand Aristotle's approach to scientific enquiry: that we can work from particular observations to deduce universal laws, and use those universal laws to make predictions about the natural world. Another English philosopher, Roger Bacon (c.1219–92), elaborated on this to develop the scientific method as we know it today: observation, hypothesis, experimentation. Bacon emphasized the importance of verification and recorded his own experiments in a way that made it possible for others to repeat his work and check his results.

From spirit to machine

The mystery of life had long been accounted for by some 'spark', energy, soul or 'breath'. In Eastern tradition, this is called *qi* or *prana*. One of the great departures of the Enlightenment involved the denial of this vitalist principle, instead seeing bodies as complex mechanisms which follow the laws of physics and operate through forces, valves, tubes and so on. This applied equally to human, animal and plant bodies. From the late 18th century, with the burgeoning of the new science of chemistry, there was much debate over whether some processes were physical or chemical. In reality, most are both.

The next big thing

The 19th century saw its own massive upheaval in the accumulating evidence of change in the natural world. Charles Darwin's account of evolution, published in 1859, changed the direction of biology forever and was the greatest development since the invention of the microscope. In its wake came genetics, and the two together – genetics and evolution – redefined the landscape of biology in the 20th century.

The big and the small

The 20th century, too, saw new interest in the interaction of organisms with one another and their environment. This has led to a broadening of scope, so that each organism is fitted into its ecological niche and seen as part of an organic and complex whole, an ecosystem that might be as tiny as the mouth of an animal or as large as the Atlantic Ocean.

Today the term 'biology' still covers the study of all living processes, from the infinitesimally small (microscopic molecular changes that control the actions of living cells) to the unimaginably vast (the way living organisms interact on a global scale). And it doesn't end with this planet. Scientists are now turning their attention, and their telescopes, to other worlds and the possibility that planets orbiting distant suns, and their moons, may also harbour life. Such organisms, if they exist, may not look like anything ever seen on Earth, but the processes which keep them alive will follow the same scientific principles that drive life on this planet and – as we shall see – have taken hundreds of years of diligent scientific enquiry to reveal.

Michelangelo's Creation of Adam *in the Sistine Chapel, Rome, shows the physical body of Adam enlivened by a divine spirit through God's touch.*

CHAPTER 1

Animal, **VEGETABLE,** mineral

'Classifying, as opposed to not classifying, has a value of its own, whatever form the classification may take. . . . Any classification is superior to chaos.'
Claude Lévi-Strauss, founder of modern anthropology, writing in 1962

Perhaps the first question ever asked in biology was 'What's that?' The urge to know what an organism is, and what it does is fundamental to our survival. Even our early ancestors, looking around them at the natural world, must have felt the need to distinguish between the huge variety of plants and animals they saw. Classification is an urge found in all cultures, however industrialized or rural.

The 9th-century Book of Animals *by al-Jahiz classified creatures in a sequence from simplest to most complex and divided them into groups based on similarities.*

Order, order

You'd think it would be easy enough to sort organisms into some kind of order. After all, there are clearly big differences between, say, a cat and a cactus, or an orange and an orangutan. But discovering, cataloguing and ordering living organisms has proved a thorny problem for more than 2,000 years.

The identification and classification of plants and animals is intellectually satisfying, helpful in structuring our thoughts, and it is practically useful, too. Which plants are edible, and which not? Which animals are dangerous and which useful? Which plants are poisonous, which medicinal, which can sting, which produce useful dyes? Which animals might eat you, which insects, spiders and snakes have a poisonous

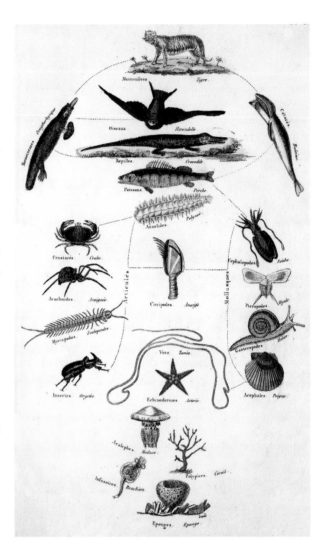

This 1834 chart of animal taxonomy bolsters the idea that mammals are nature's peak achievement.

MANY WAYS

The anthropologists of the mid-20th century collected many instances of tribal groups with discrete names for a vast array of animals and plants found in their environments. The American Robert Fox, writing of the Pinatubo Negritos of Southeast Asia in the 1950s, reported that they had names for at least 450 kinds of plant, 75 birds, and even 20 species of ant. Some of these were of direct use to the Negritos, but by no means all. Others had an impact on plants or animals with which the Negritos were concerned.

– or just painful – bite or sting, and which animals run too fast to catch? To share and preserve such knowledge it is necessary to name and describe natural organisms. But biology goes beyond classifying organisms according to their relationship to humans as edible, dangerous and so on. It involves discovering (or imposing) a structure that relates organisms to one another.

Paleolithic zoology

Before the earliest writing, humans created cave art which depicts the animals they saw and hunted. These illustrations are often detailed enough for modern zoologists to identify and learn something about the animals from this time.

In 2011, the French evolutionary geneticist Melanie Pruvost found that the colours of horses shown in Paleolithic art from Siberia, East and West Europe and the Spanish peninsula matched fossil DNA

evidence. It is even possible to learn new information about some of the animals depicted from the coloration and patterns used by early artists. Spotted horses had not been thought to exist in Europe at the time they are shown in paintings, but new DNA evidence supports the testimony of the cave art.

The variety of species depicted in some cave art makes the paintings a veritable gallery of prehistoric fauna, and it's possible that one of its uses was to help young hunters identify different types of animal. Most images are of adult animals viewed in profile, which makes their identification easier than if drawn from an oblique angle (it's also the easiest way to draw them!)

More than a backdrop

Once people began to live in settled communities, they recorded more complex interactions with the natural world. In the backdrop of a relief panel from the ancient Assyrian city of Nineveh, carved around 700BC, deer are shown among the giant reeds *Phragmites australis*, which were used for fuel and animal fodder and to make boats and mats. One relief shows lions under a pine tree twined with grape vines, and

Cave paintings of animals, such as this in Chauvet, France, are often astonishingly accurate.

13

another shows many recognizable species of trees growing in a park. A wall-painting in Santorini, Greece, dating from 1500BC, shows women collecting the saffron crocus (*Crocus sativus*), and another shows red lilies (*Lilium chalcedonicum*) both in bud and in full flower. That these depictions are sufficiently accurate for modern biologists to identify the species shows that the artists had a genuine interest and a keen eye – they were early biological illustrators, alert to the differences between organisms.

Depicting the useful

Early depiction and classification of plants and animals was probably driven by utility. Those animals or plants that were useful to people were of most interest: they could be eaten, farmed, skinned, used in medicines or put to work. Sometimes they had religious significance or simply played an important role in people's lives. In Ancient Egypt, the scarab beetle was regarded as sacred and is often depicted in meticulous detail. The scarab is a dung beetle and its vital role in removing the doubtless large volumes of

dung produced by humans, camels and other animals cannot have gone unnoticed in a region where there was scarce rain to wash the muck away. The ibis, too, had a religious significance and was often depicted. It might seem less immediately useful, but ibis helped to remove snails from fishponds, and the snails often carried dangerous liver parasites. There was no need to understand how, but if the presence of ibis near the fishponds reduced the incidence of parasites in the people who ate the fish, the bird would get credit for its work.

The scarab beetle was revered in Ancient Egypt and often depicted on jewellery.

Classifying life

The naming and classification of organisms is known as taxonomy. Biological taxonomy began in Ancient Greece, with the work of Aristotle (384–322BC). He was the first author to attempt a system of classification based on intrinsic characteristics in the organisms, rather than their usefulness to humans.

For Aristotle, the goal of science was to make something systematic and coherent out of the disorder that is our observation of the world. His writings on biology were part of his greater investigation into the nature of knowledge and how it can be acquired. He saw four questions to be answered in any enquiry, and they fall into pairs:

'We inquire about four things: the fact, the reason why, if something is, what something is . . .

When we know the fact we inquire about the reason why . . .

And having come to know that it is, we inquire what it is.'
Aristotle, *Posterior Analytics*

SEE-THROUGH PHARMACIST

The first attempt at the systematic identification and naming of plants was recorded in China around 2,000 years ago. It was a pharmacopoeia called *The Divine Farmer's Herb-Root Classic*, compiled in the Western Han Dynasty, 206BC–AD220, but now lost. Legend tells that it was written by Shennong, or the 'emperor of the five grains', who was conveniently transparent, making it easy to see the effects of his medicinal experiments on himself. (If Shennong lived at all it was in the 24th century BC, around 2,000 years before the text was written.) The book apparently named 365 medicines derived from plants, animals and minerals, but because it focused on the medicinal products rather than the organisms themselves it is not properly a work of taxonomy.

Dividing animals

The natural world does not present itself to us in an ordered way, with the relationships between plants and animals clearly laid out. So Aristotle approached the natural realm with the intention of finding order. He saw three types of 'thing' in the natural world: animals, plants and minerals. The first two categories have come to be the subject matter of biology.

Aristotle set out first to distinguish between animals in his *History of Animals* and then to explain the reasons for the observed features in *On the Parts of Animals*.

His task began, then, with classification. And the task of classification, in turn, began with rules for classifying.

Aristotle warns against arbitrary categorization, saying that we must retain a consistent thread of relatedness throughout a system. For example, if we divide animals into wild and tame, and then divide the tame animals into light-coloured and dark-coloured, that is a nonsensical hierarchy, as colour does not relate to their tameness or wildness. He argues for high-level grouping that reflects things being 'alike in kind' – so all fish are alike in a way that birds are not like fish. Within the group of birds, there will be birds with long beaks and birds with short beaks; that is a division in degree (so a division in 'birdness'). To fall into a group, animals must have 'common natures, and forms not very far apart'.

Tab XXV

The *History of Animals* isn't concerned with cause and effect, so it remarks that all animals that have blood have a heart, but does not suggest that they *must* have a heart or *why* they must have one – just that they do. *On the Parts of Animals*, on the other hand, attempts to explain why things are as they are. For example, Aristotle

Lacking lungs, fish have no need of a neck and larynx.

finds that animals with lungs have a neck. This is because the larynx is needed for the air to travel to the lungs and to separate into the branches going to each of them. Animals with a neck have an oesophagus, but not because it is needed for nutrition – only because it is needed to carry the food from the mouth and through the neck to the stomach: the presence of the neck necessitates the oesophagus. So animals without lungs (such as fish) have no neck and no oesophagus.

Genus and species

Aristotle introduced a two-part naming system, which is the predecessor of the one in use today. He named each specimen with a generic name (genus) which identifies the family or race of the organism and then a descriptive part (species) which defines its difference from others of the same genus. The 'difference' part aimed to be sufficiently descriptive that the organism could be distinguished from others by referring only to the name. This worked reasonably well when applied to the local flora and fauna of the Mediterranean area, but faltered as more and more organisms were discovered hundreds of years later (*see* page 31).

The *History of Animals* gives a full descriptive account of how animals are similar and how they vary, by reference to their structure, organs, tissues, methods of reproduction, habits, means of locomotion and so on. A passage on feet gives a feel for how this goes:

'Of blooded and viviparous quadrupeds some have the foot cloven into many parts, as is the case with the hands and feet of man (for some animals, by the way, are many-toed, as the lion, the dog, and the pard); others have feet cloven in twain, and instead of nails have hooves, as the sheep, the goat, the deer, and the hippopotamus; others are uncloven of foot, such for instance as the solid-hooved animals, the horse and the mule.'

History of Animals, book 2

(Aristotle's distinction between 'blooded' and 'unblooded' animals is equivalent to our distinction between vertebrates and invertebrates.)

WHICH CAME FIRST, THE CROCODILE OR THE CROCODILE'S TAIL?

Aristotle saw the body structure and processes of an animal as being those necessary to perform the functions that the form (soul, essence or nature) of the animal demands. This means, for instance, that the crocodile has a powerful tail because it needs to move through water quickly in order to be a crocodile. It is the opposite of saying a crocodile can move through water quickly because it has a powerful tail. Which came first, the tail or the crocodile? For Aristotle, the crocodile came first and necessitated the tail, rather than the tail facilitating the crocodile. It might not look like an important distinction, but it will become important later when evolution appears on the scene (*see* pages 36–7).

Aristotle treats living things in the order of their importance (as he saw it) with the most important first, so starting with humankind, then blooded animals, then unblooded animals. He uses methods of generation (reproduction) to put animals into a hierarchy, with those that give birth to fully formed young at the top and those he considered were spontaneously generated (springing from inanimate matter such as mud) at the bottom. (The same distinction occurs in the Ayurvedic tradition in India, dating from around 1500BC.) In the final book, Aristotle deals with enmity and cooperation between animals, and with their perceived characters:

'The characters of animals, as has been observed, differ in respect to timidity, to gentleness, to courage, to tameness, to intelligence, and to stupidity.'

In all, Aristotle introduces some general principles of categorizing animals – that we must look for general groups and then find distinctive, relevant differences between the members of a group. But he ends with a descriptive catalogue rather than a rigorous system of classification. Still, he set an important precedent. Cataloguing was to remain the dominant trend in biological writing for around 2,000 years.

From animals to plants

What Aristotle did for animals, his successor Theophrastus (c.371–287BC) did for plants, in *Enquiry into Plants* and *On the Causes of Plants*. Theophrastus divided plants into domesticated trees, wild trees, shrubs and herbaceous plants. He acknowledged, though, that distinctions such as 'wild or cultivated', and 'bush or tree' are not very rigorous, and that classifying plants is difficult – the categories overlap. The division between aquatic and terrestrial plants is different as it is natural and objective – plants either do or do not live in the water.

Theophrastus was aware that it is harder to speak in general about plants than about animals because there are no parts of plants that are common to all. While all animals have, say, a mouth, there is nothing – not stalk, nor leaves, nor flowers nor roots – that all plants have. He discovered, too, that some wild plants differ from their cultivated forms, even though they are apparently the same type. He focuses, therefore, on giving accurate

Aristotle's thoughts on animals and how we should treat them are still influential today.

accounts based on observation rather than trying to use reason to draw conclusions about plant life.

Theophrastus gives a large amount of detail on different species of plants, tackling 550 plants from an area stretching from the Atlantic, around the Mediterranean and as far as India. He travelled through Greece researching plants, kept his own garden, took advice from professionals and experts, and examined plants brought back from military expeditions.

Theophrastus wrote the earliest known text on the classification of plants.

first encyclopaedia. Pliny's text deals with all kinds of topics, including painting, ethnography and geology, but books 8–11 cover zoology and books 12–27 deal with botany (including horticulture and medicine from plants). Pliny was an enthusiastic scholar who devoted every spare moment to research and writing. He died in the eruption of Vesuvius at Pompeii in AD79, keen to investigate even that terrifying phenomenon, according to his nephew, Pliny the Younger. His zoological writing records all the accounts he could find of animals of every kind and from every known region. Not surprisingly, there are many inaccuracies, with mythical animals recorded as genuine and some very unlikely characteristics noted without comment.

The first encyclopaedias

Animals and plants are found together in the work of the Roman writer Pliny the Elder. In the 1st century AD, he composed his *Natural History*, a collection of 37 volumes which aimed to record everything known about the natural world – essentially, the

'The mantichora . . . has a triple row of teeth meeting like the teeth of a comb, the face and ears of a human being, grey eyes, a blood-red colour, a lion's body, inflicting stings with its tail in the manner of a scorpion, with a voice like the sound of a pan-pipe blended with a trumpet, of great speed, with a special appetite for human flesh. . . .

'The basilisk serpent . . . is a native of the province of Cyrenaica, not more than 12 inches [30cm] long, and adorned with a bright white marking on the head like a sort of diadem. It routs all snakes with its hiss, and does not move its body forward in manifold coils like the other snakes but advancing with its middle raised high. It kills bushes not only by its touch but also by its breath, scorches up the grass and bursts rocks. Its effect on other animals is disastrous: it is believed that once one was killed with a spear by a man on horseback and the infection rising through the spear killed not only the rider but also the horse.'

Pliny the Elder, *Natural History*, book 8, AD77–9

What's in a name?

The work of Pliny became the foundation of the many medieval encyclopaedias, beginning with the historian Isidore of Seville's *Etymologies* and *On the Nature of Things*. The encyclopaedias focused on the usefulness to humankind of the animals, plants and minerals described in their pages, taking a highly teleological and anthropocentric view: everything in nature is there for a purpose, and it's a purpose that is intended for human exploitation.

Isidore of Seville (*c.*560–636) was the first Christian encyclopaedist and perhaps the last Classical one. In the *Etymologies*, he demonstrated his belief that the name of a thing – or the origin of the name of a thing – reveals

Above: Pliny's Natural History *remained popular throughout the Middle Ages, as is clear from this lavish 15ᵗʰ-century manuscript copy.*
Left: Ants carrying a particle of food.

the thing's true nature. So, for example, he explained that the ant is called *formica* (in Latin) because it is *formis* (strong) and carries *mica* (particles).

The encyclopaedic tradition continued well into the Middle Ages, with no one pausing to doubt the information handed down from antiquity. Reverence for the

> 'Adam first named all living creatures, assigning a name to each in accordance with its purpose at that time, in view of the nature it was to be subject to. But the nations have named all animals in their own languages. But Adam did not give those names in the language of the Greeks or Romans or any barbaric people, but in that one of all languages which existed before the flood, and is called Hebrew.'
>
> Isidore of Seville, *Etymologies*, AD600–25

Classical world, and the general belief that the world was becoming increasingly imperfect, meant that the writers of antiquity were generally considered to be more reliable than any contemporary source and rarely challenged. Instead, they were generally venerated and mined for ever-deeper levels of meaning.

A comedy of errors

While Pliny intended to provide a factual account of the nature of animals and plants, the Greek text known as *Physiologus*, probably originating in the 3rd or 4th century in Alexandria, aimed instead to uncover the deep, allegorical, spiritual meanings embedded in nature. These meanings were supposedly put there by God in his design for Creation, making the natural world a kind of book in which we can learn something of the divine purpose, if only we know how to read it.

Starting in the 12th century, the encyclopaedias and the *Physiologus* came together in the medieval bestiaries, which combined supposedly accurate zoological knowledge with an account of the allegorical 'meanings' of animals. There had been virtually no progress in terms of biological knowledge over the 600 years that separate the early encyclopaedists from the first

bestiaries. The prevailing view remained that the whole of the natural world was created with a purpose, and the overall purpose was to manifest God's power and to serve humankind. The impetus for observing an animal or its behaviour was

Depictions of a mole and the mythical leucrota (a cross between a hyena and a lioness) from the 13th-century Northumberland Bestiary.

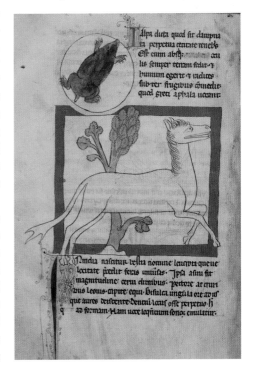

21

> 'Let us make man in our image, after our likeness: and let them have dominion over the fish of the sea, and over the fowl of the air, and over the cattle, and over all the earth, and over every creeping thing that creepeth upon the earth.'
>
> Genesis, chapter 1 verse 26, King James Bible, 1611

to find the meaning hidden in it, not to understand the functioning or ecological place of the animal.

It would be difficult to overstate the degree of anthropocentric thinking that characterized the European Middle Ages. For Roger Bacon (c.1214–c.1292), humankind was the lynchpin of the natural order; take humans away, and it all becomes purposeless chaos:

> 'Man, if we look to final causes, may be regarded as the centre of the world; inasmuch that if man were taken away from the world, the rest would seem to be all astray, without aim or purpose.'

Medieval bestiaries always looked for a lesson for humankind in the way God had organized the natural world. So, for example, the ostrich looks up to the heavens, then lays its eggs in the sand and walks away from them. This instructs us to look up to God in heaven and disregard worldly things. But for all their good intentions

Hildegard of Bingen's Physica, *written in the 12th century, is a catalogue of the medicinal properties of stones, plants, fish, reptiles and other animals, all created by God for human use.*

GEESE FROM BARNACLES

'They are produced from fir timber tossed along the sea, and are at first like gum. Afterwards they hang down by their beaks as if they were a seaweed attached to the timber, and are surrounded by shells in order to grow more freely. Having thus in process of time been clothed with a strong coat of feathers, they either fall into the water or fly freely away into the air. They derived their food and growth from the sap of the wood or from the sea, by a secret and most wonderful process of alimentation. I have frequently seen, with my own eyes, more than a thousand of these small bodies of birds, hanging down on the sea-shore from one piece of timber, enclosed in their shells, and already formed. They do not breed and lay eggs like other birds, nor do they ever hatch any eggs, nor do they seem to build nests in any corner of the earth.'

Giraldus Cambrensis, *Topographica Hiberniae*, 1187

Ripe barnacle geese drop into the water from the branches of a tree on which they have grown.

to uncover the meaning of the natural world, the bestiaries are a treasure-trove of endearing biological errors. It would be easy to be condescending, but to do so would be to lose an insight into the scientific reasoning behind them. The authors did try to explain observations as well as give a spiritual explanation for the way the natural world is. An example is the account of the barnacle goose.

The barnacle goose was never seen to nest, and with good reason – it is a seasonal visitor to most of Europe and is not seen in the summer while raising its brood in the Arctic. The account of the barnacle goose given in the bestiaries (and repeated well into the 16th century) was that the goose grew on trees or on pieces of wood, then dropped into the water where it continued to develop as a barnacle, and finally emerged as a goose, appearing from over the sea in the autumn. Some people believed this theory allowed the consumption of barnacle geese during Lent and on fast days, since they could not really be meat if they grew on trees. (Though the Fourth Lateran Council ruled against eating barnacle geese in 1215.)

23

The Great Chain of Being

As we have seen, Aristotle not only divided animals into classes but proposed a hierarchy of living things, the *scala natura* (ladder of nature). Humans were at the top of this hierarchy, followed by 'blooded' animals (vertebrates), 'unblooded' animals (invertebrates), and then plants. Aristotle believed each organism has a soul of the type needed for its capabilities – in fact, the soul is what bestows the capabilities. Each organism also has a body structure suited to its capabilities, as with the crocodile and its tail. So a human has a superior soul capable of reason, locomotion, growth and sustaining life; animals can move, grow and sustain life; but plants can only grow and sustain life. The idea of a hierarchy remained the dominant model for thinking about the natural world and endured until the 19th century, providing a structure for far more than natural history. Indeed, it encompassed social, political and even divine realms.

The bestiaries declared that baby bears are born formless and the mother licks them into shape – a belief that may have come from seeing bears lick the amniotic sac from around their infants.

In the 3rd century AD, the Neoplatonists developed the *scala natura*, adding new rungs to accommodate the gods above humankind. In the late 5th century, the Neoplatonist philosopher Pseudo-Dionysius the Aeropagite Christianized the system, replacing the pagan gods on

> 'All things, however different, are linked together. There is in the genera of things such a connection between the higher and the lower that they meet in a common point; such an order obtains among species that the highest species of one genus coincides with the lowest of the next higher genus, in order that the universe may be one.'
>
> Nicolas of Cusa (1401–64)

the top rungs with the angels and the Christian God. Philosophers in the Middle Ages continued to develop and endorse the *scala natura*, emphasizing the Christian and social implications of seeing the world order in terms of a hierarchy. One of the most influential proponents was Italian philosopher Thomas Aquinas (1225–74), who put the divine beings into a precise order, with seraphim as the most important rank of angels. Significantly, Aquinas was responsible for the integration of Aristotle into Christian thought.

From the Middle Ages onwards, the ladder of nature was more often referred to as a great 'chain of being'. It had God at the top, followed by angels, humans, animals, plants and inanimate matter such as metals and stones at the bottom. The hierarchy could be broken down minutely; so within plants there is a hierarchy, and there is even one within stones and metals. Things further up the ladder had more 'spirit' and less 'matter' – so angels, being all spirit, were

closest to God. Lead, being very heavy, has a lot of matter so was low even in the order of metals. Alchemy, incidentally, sought to change base metals into gold by adding 'spirit'; gold was believed to have more spirit than other metals.

The Great Chain of Being, depicted here in 1579, puts God at the top, angels below, and then humans at the head of the chain of earthly creatures.

ALL IN ORDER

The typical sequence of the chain of being from the Middle Ages, missing out angels as they are not really subjects of biology, was as follows. (Within each group, the first listed was considered the 'prime' or 'principal' of that class of being.)

HUMANS – a special type, with some of the rational and spiritual power of angels but, unlike angels, tied to a physical body. MAMMALS – with elephant or lion as the principal:
- Wild animals
- 'Useful' domesticated animals (e.g. horse, dog)
- 'Tame' domesticated animals (e.g. cat)

BIRDS – with the eagle as the principal. Birds were higher than aquatic animals as the element of air was considered superior to water:
- Birds of prey
- Carrion-eating birds
- Worm-eating birds
- Seed-eating birds

AQUATIC CREATURES, with whale as the principal:
- Aquatic mammals
- Sharks
- Mobile fish
- Immobile shellfish

PLANTS, with oak as the principal:
- Trees
- Shrubs
- Bushes
- Crop plants
- Herbs
- Ferns
- Weeds
- Moss
- Fungus

MINERALS, with diamond as the principal:
- Gems
- Metals (gold is the principal metal)
- Geological rocks (marble is the principal)
- Minute particles (sand, soil, etc.)

'Infinite variety'

The chain was considered to be a continuum with many tiny intermediate steps. For example, shellfish were low in the scheme of animals, forging a link with plants, as they move little or not at all. The chain of being has no gaps – no 'missing links'. By the time the German scholar Albertus Magnus presented his version of Aristotle's work on natural history in the 13th century, many more organisms had been discovered, but the links of the chain could be prised apart to accommodate them.

The law of continuity satisfied the idea that the world is perfect, being filled with every possible type of organism, nothing missed out. It is both complete and unified. This concept of completeness explained why even seemingly inferior or 'useless' organisms such as mosquitoes or maggots are part of Creation. The German philosopher and mathematician Gottfried

'Nature does not make [animal] kinds separate without making something intermediate between them; for nature does not pass from extreme to extreme without an intermediate.'

Albertus Magnus, *De animalibus*, 1450–1500

'All the different classes of beings which taken together make up the universe are, in the ideas of God who knows distinctly their essential gradations, only so many ordinates of a single curve so closely united that it would be impossible to place others between any two of them, since that would imply disorder and imperfection. Thus men are linked with the animals, these with the plants and these with the fossils, which in turn merge with those bodies which our senses and our imagination represent to us as the absolutely inanimate . . . all the orders of natural beings form but a single chain, in which the various classes, like so many rings, are so closely linked one to another that it is impossible for the senses or the imagination to determine precisely the point at which one ends and the next begins.'

Gottfried von Leibniz, 1753

Gottfried von Leibniz, champion of a complete and perfect Creation.

von Leibniz (1646–1716) was particularly famous for his philosophy that the world is the best it could possibly be, given the requirement for God to have created all that He could have created. If the world is to be complete, it must include evil, and parasites. (It was Leibniz's position on 'the best of all possible worlds' that Voltaire satirized so scathingly in *Candide* in 1759).

Everything in its place

It would be disingenuous to suggest that the chain of being was solely or primarily a model of biological taxonomy. It was also theological and political; it formed the framework for an entire world-view of which the natural world was an important part. It was right for the monarch to rule over the people, for the serf to be subservient to the landowner, for children to obey parents and for women to be subordinate to men. This model served the rulers very well indeed, and we can imagine they would have been keen to keep it in place.

The origins of zoology

The 16th century saw a revolution in the sciences. Suddenly, the wisdom of the ancients was being challenged in many spheres. Copernicus (1473–1543) had overthrown the Ptolemaic view of the solar system that put Earth at the centre and instead put the Sun in its rightful place. The Italian explorer Amerigo Vespucci (1454–1512) had demonstrated that the Americas were a new, previously unknown, continental landmass, not part of Asia; and Andreas Vesalius (1514–64) had shown that Galen (*see* page 49) was wrong about many aspects of the human body. Against this background, a Swiss physician, botanist and natural historian, Conrad Gessner (1516–65), brought biology into the modern era.

Nature's plenty

Gessner forms a bridge between the ancient and the modern approaches to investigating nature. He wrote what is considered the first zoological text, a comprehensive account of all known animals which made a serious attempt at scientific rigour. The first four volumes of his

Gessner's drawing of a marmot, Marmota, *a type of large squirrel.*

Historiae animalium, filling 4,500 pages, appeared in 1551–8, covering mammals (live-bearing quadrupeds), amphibians (egg-laying quadrupeds), birds and fishes; a fifth volume on snakes appeared after his death, in 1587. He gives extensive descriptions of the animals, with accounts of their habits and behaviour alongside their uses (as food, medicine, and so on). His was the first book to attempt to illustrate the animals in their natural environment, and he was also the first to depict fossils.

Gessner still took material from traditional authorities including the Old Testament, Aristotle, Pliny and the bestiaries, and still sought to elucidate the divine messages hidden in the natural world. But he supplemented this traditional content with his own observations, setting out to describe animals accurately. Guided by four principles – observation, dissection, travel, and accurate description – he attempted

a complete survey of the animal kingdom. He drew on advice and received samples from many of the naturalists of his day, and had an extensive network of sympathetic contacts. Mythical animals such as the unicorn and mermaid are still there, but he did express his doubts when he was uncertain whether an animal existed as described.

Biological treasure troves

Among the wealthy and the nobility it became fashionable in the 16th century to keep a 'cabinet of curiosities' or *Wunderkammer*. This was a collection, kept in a dedicated room, of all kinds of natural and manufactured wonders with a distinct bias towards natural history. Typical contents included stuffed animals and fish, horns (especially narwhal horns, often recorded as unicorn horns) coral, skeletons, fossils, strange plants (including the vegetable lamb), drawings of deformed animals and people, and even preserved foetuses.

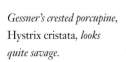

Gessner's crested porcupine, Hystrix cristata, *looks quite savage.*

Ole Worm's cabinet of curiosities, 1654.

These were all kept alongside sculptures, archaeological finds, automata, odd minerals and anything else that looked interesting. Among the most famous *Wunderkammer* were those of Rudolf II (Holy Roman Emperor, 1576–1612), Ole Worm (1588–1654) and Athanasius Kircher (1602–80).

The aim was to assemble a heterogeneous collection of wonders that celebrated the diversity of the world and served, in effect, as a fantastical microcosm. There was little attempt to classify the items, but rather to celebrate their richness and variety. Even so, some scientific advances came from the distribution of images of items, which might then be compared with other examples and identified.

Some people went further and collected menageries of unusual animals. The animals, sadly, were generally not well or suitably cared for and usually came to grief. The same enthusiasm was extended to plants, with some botanists travelling great distances to collect plants for their own gardens or those of their patrons. John Tradescant the elder (1570–1638) and the younger (1608–62) travelled in Europe and America to 'gather up all raritye of flowers, plants, shells, &c.' for the gardens of King Charles II. The Tradescants were the first to acknowledge the value of scientific collections to the public and opened theirs to paying visitors, housed in the 'Ark' in Lambeth, London. On the death of the younger Tradescant, the collection passed to Elias Ashmole and became the heart of his cabinet of curiosities. In 1677, he donated it to Oxford University and the Ashmolean Museum was founded to house it. It can still be viewed there.

'In the museum of Mr. John Tradescant are the following things: first in the courtyard there lie two ribs of a whale . . . all kinds of foreign plants. . . . In the museum itself we saw a salamander, a chameleon, a pelican, a remora, a lanhado from Africa, a white partridge, a goose which has grown in Scotland on a tree, a flying squirrel, another squirrel like a fish, all kinds of bright colored birds from India, a number of things changed into stone.'
Georg Christoph Stirn, 1638 – extract from his travel diary

One of the most famous zoological drawings, Dürer's picture of a rhinoceros, illustrated Gessner's Historiae animalium. *Sent by Manuel I of Portugal to Pope Leo X, the rhino died in a shipwreck off the Italian coast in 1516.*

Returning to life

Instead of relying on ancient reports, 17th-century naturalists wishing to catalogue or write about plants and animals began to collect their own specimens, not simply in the haphazard acquisitiveness of the *Wunderkammer*, but with a new spirit of rigorous enquiry in keeping with the scientific method.

One of the first and most influential naturalists in this tradition was the English clergyman John Ray (1627–1705). In 1663, Ray set out with the English ornithologist and ichthyologist Francis Willughby and two others to travel around Europe. They returned in 1666 with a large number of biological specimens that they intended to describe and catalogue, drawing up a systematic means of classification for the natural world. But Willughby died in 1672, leaving work on birds and fish for Ray to edit. Ray wrote extensively on plants, but their grand scheme was left unfinished. Even so, Ray was the first to give the formal definition of 'species' (*see* box below) and his catalogue of British plants published in 1670 provided the basis for later volumes on English flora.

John Ray was one of the first English parson-naturalists.

'No surer criterion for determining species has occurred to me than the distinguishing features that perpetuate themselves in propagation from seed. Thus, no matter what variations occur in the individuals or the species, if they spring from the seed of one and the same plant, they are accidental variations and not such as to distinguish a species . . . Animals likewise that differ specifically preserve their distinct species permanently; one species never springs from the seed of another nor vice versa.'

John Ray, *History of Plants* (1686)

Naming names

What Ray might have done was completed a century later by the Swedish botanist Carl Linnaeus (1707–78). Instead of just gathering vast numbers of plants, as the Tradescants and subsequent collectors had done, Linnaeus set out to catalogue all the plants he could trace, finding the similarities and differences between them and identifying the species. In 1749, he devised the two-part nomenclature that is still used to designate the genus and species of an organism.

Linnaeus worked primarily on plants, and had begun by exploring the new claim that plants exhibit sexuality. He divided flowering plants according to the form of their stamens and subdivided them according to the number of pistils each had. This was of limited use, but spurred Linnaeus on to an extendable system.

Next, he gathered together all the species and grouped them into related genera. He then introduced a system that gave a noun for the genus followed by an adjective for the species. So, for example, when we get to animals, we have *Panthera*

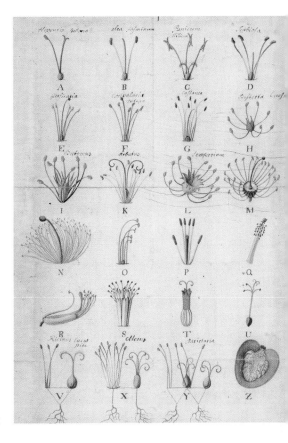

Linnaeus divided plants into 23 classes according to the number and arrangement of the flowers' sexual organs.

Linnaeus described nine stamens and a pistil as 'nine men in the same bridal chamber with one woman'.

pardus (leopard) – *Panthera* is the genus (big cats) and *pardus* the species (the spotty ones that are leopards). This replaced the Aristotelian system, which had become very cumbersome as more and more species were discovered. Since Aristotle required that the second part of the name, the 'difference' part, was sufficiently descriptive to distinguish an organism from all others, it was often long – and grew longer as more and more things had to be uniquely distinguished. The name for the tomato, first encountered by Europeans in the 16th century, was *Solanum caule inerme herbaceo, foliis pinnatis incises, raconis simplicibus*, which means 'Solanum with a smooth herbaceous stem, incised pinnate leaves and simple inflorescence'. The Linnaean name is *Solanum lycopersicum* – much more concise!

Linnaeus firmly believed that once all the organisms had been classified, the job of taxonomy would be done forever – he did not think organisms changed, but that everything was fixed at Creation and had remained the same since. Although new organisms were discovered as people travelled or looked through a microscope, no new ones would ever arise in nature. He saw his task as a biologist as completing the work of Adam in naming the plants and animals and marvelling at God's creation.

However, despite his conventionally religious stance, Linnaeus was the first person to suggest that humans and primates are comparable, and to treat humans as another type of animal. For him, there was a difference of degree, but in essence humans were not sufficiently distinct from other animals to be excluded from his attention.

> 'The first step in wisdom is to know the things themselves . . . objects are distinguished and known by classifying them methodically and giving them appropriate names . . . classification and name-giving will be the foundation of our science.'
> Carl Linnaeus, *Systema Natura* (1735)

Class system

The Linnaean system has a hierarchical structure, but does not suggest that one organism is in some way 'better' or 'higher' than another. Each belongs to one of two kingdoms of nature (plants or animals), and organisms are then divided by class, order,

CARL LINNAEUS (1707–78)

Born in Sweden to a father who was a pastor and keen gardener, Linnaeus began to train as a physician but had to give up when he ran out of money. He moved to Uppsala where he worked in the botany department of the university, taking over a lecture course on plants. His inspirational teaching increased the audience for the lectures from around 80 to 400. In the 1730s he travelled extensively in Europe collecting plant samples. In 1735 he went to Holland and finally trained as a physician, also publishing his taxonomy of flowering plants. He returned to Sweden as a qualified doctor and continued to work on botany. His fascination with classification and his skill with people drew others into his work. A large proportion of his many devoted students travelled in search of new specimens – and up to a third of them possibly died in the process.

genus and species. Linnaeus listed minerals as a third kingdom, following Aristotle's early division, but did not consider them to be living things. Linnaeus recognized six classes:

- Mammalia (mammals)
- Aves (birds); Linnaeus was the first to classify bats as mammals rather than birds
- Amphibia (amphibians, reptiles, and some non-bony fish)
- Pisces (bony fish), including spiny-finned fish (Perciformes) as a separate order; whales and manatees were originally classed as fish
- Insecta (all arthropods); crustaceans, arachnids and myriapods formed the order Aptera
- Vermes (remaining invertebrates, roughly divided into worms, molluscs, echinoderms and other hard-shelled invertebrates).

There were innumerable orders in each class, which divide animals into groups such as *Crocodilia*, which includes all the crocodile-like animals (crocodiles, caymans, alligators and so on). Within the order, the genus specifies a precise type of animal – such as crocodile (*Crocodylus*) rather than alligator (*Alligator*) – and the species indicates the precise group which can interbreed (*Crodocylus niloticus*, Nile crocodile). Modern biologists recognize eight classes.

Breaking the chain

The Great Chain of Being had proved remarkably resilient. The age of exploration brought new discoveries, from North and South America, hosts of islands (the Caribbean, various Pacific Islands, those in Southeast Asia) and finally Australasia, showing that the 'complete' creation was far

Linnaeus divided animals into six classes and a group of 'paradoxa' that included the rhinoceros, pelican and phoenix.

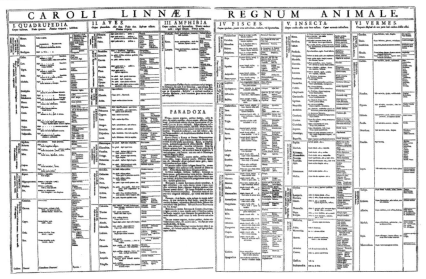

LINNAEUS AND THE HYDRA

While in Hamburg, Linnaeus was shown a curiosity owned by the mayor. Known as the 'Hamburg hydra', it was supposedly a stuffed, seven-headed creature. Linnaeus quickly established it was a fake made from stitched-together snakeskins with the feet and jaws of weasels. He not only revealed it as a fake, but published the revelation, angering the mayor who had hoped to sell it for a large sum!

Albertus Seba illustrated the Hamburg hydra in 1734.

larger than previously thought. From the 17th century, the invention of the microscope showed another whole world of tiny beings (*see* pages 86–105). Yet all these could be found a place in the chain. There were, after all, spaces where no one thought there had been. It was not living creatures that would undo the chain.

> '[Microorganisms] fill the space nature has left between the simple, living, organic molecule on the one hand, and animals and vegetables on the other. This sequence, this chain of being descending from the most highly organized animal to the simple organic molecule, admits of all possible degrees, all imaginable nuances.'
>
> Georges-Louis Leclerc, Comte de Buffon (1777)

No spaces, no change

God's creation was considered not only complete, but unchanging. If everything is in its appointed place in a hierarchy, if all that can exist does exist, there is no room for change. Newly discovered organisms were fitted into the chain with little difficulty, but eventually evidence of change began to challenge this 2,000-year-old model of the natural hierarchy.

Gessner published the first illustrations of fossils in *Historiae animalium*. In 1565, he published the first treatise on palaeontology, *De rerum fossilium* (*On fossil objects*) in which he gathered together all the fossil evidence he could find, drawing on help from others around Europe. He intended it to be the start of a great work on fossils, but unfortunately died of the plague before he could do any more. Some of the fossils he described he was able to identify as similar to extant creatures, but others presented more of a puzzle. He put several ammonite fossils alongside gastropods and snakes – an inevitable consequence of being unable to countenance animals that had once lived but could no longer be found.

It was a missed opportunity, but perhaps one the world was not ready for. Two hundred years later the world was ready, just, when Georges-Louis Leclerc, Comte de Buffon (1707–88) published the first volume of his ambitious (and uncompleted) *Histoire naturelle*. This aimed to be an encyclopaedic account of all animals, plants and minerals. Leclerc scandalized Paris with his contention that the Earth was much older than its generally accepted age (dating from 4004BC) and that animals had changed over the course of time. His heretical views were condemned by the Sorbonne, and Leclerc publicly retracted them – but he continued to publish without alterations. Noting that similar but not identical species exist in different parts of the world, he suggested that all animals had spread out from a single location and changed over time to adapt to local conditions. He also suggested that climate change might have influenced the development of animals and plants.

However unpopular Leclerc's ideas were with the establishment, they were influential. One of those who followed

COLLECTING BIRDS

The quest to find and catalogue continued well into the 19th century. In about 1820, the American naturalist and artist John James Audubon (1785–1851) set out to paint every one of the native birds of America. The result was *Birds of America*, a beautiful large-format book that ran to 435 colour plates. It was sold unbound to avoid the legal obligation of depositing copies in UK libraries (Audubon was living in the UK at the time). Later editions were smaller, bound and cheaper. Copies of the first edition change hands occasionally, commanding around US$10 million.

Audubon's ruffed grouse.

in his footsteps was the French naturalist Georges Cuvier (1769–1832). Acclaimed the founder of vertebrate palaeontology, he was the first to provide compelling evidence of extinction. He showed that large animals such as mammoths and giant ground sloths, which are preserved as fossils, cannot be identified with any living creatures. And they are too large to have been missed. He presented his devastating findings in 1796, when he was just 26 years old. In 1812 he provided incontestable proof of change over time, with older fossils in lower rock strata and newer fossils in strata close to the surface. Although he could not – and did not try to – explain how change happened, he established once and for all that animals can and do change over time. The fixed chain of being was no longer a tenable model of the natural world.

From chain to tree

There is a single illustration in Charles Darwin's groundbreaking work on evolution, *On the Origin of Species by Natural Selection* (*see* page 149), and it is of a 'tree of life'. The illustration first appears as a rough sketch in one of Darwin's notebooks (*see* opposite). It is the point at which taxonomy and evolution meet. The idea is that as later organisms have evolved from earlier organisms, we can trace the relationships between them, just as we can trace relationships within a family tree. Darwin had no way of testing theories about the evolutionary relationships between organisms, and still had to rely on morphology (form) and behaviour to suggest them.

Instead of a parade of organisms, fixed since the time of God's Creation, which Linnaeus believed he was dealing with, Darwin and others who accepted evolution saw the contemporary natural world as a snapshot of a process. Behind it lay organisms now extinct, but which could, through their very existence, explain why a whale might look more like a fish but actually be more like a cow. Instead of

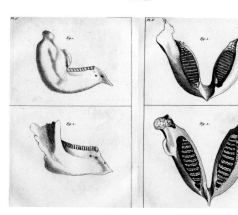

Cuvier compared the jaws of an elephant and a mammoth in a paper published in 1798–9.

'As buds give rise by growth to fresh buds, and these, if vigorous, branch out and overtop on all sides many a feebler branch, so by generation I believe it has been with the great Tree of Life, which fills with its dead and broken branches the crust of the earth, and covers the surface with its ever-branching and beautiful ramifications.'

Charles Darwin, *On the Origin of Species*, 6th edition, 1872

Darwin's rough sketch of the Tree of Life.

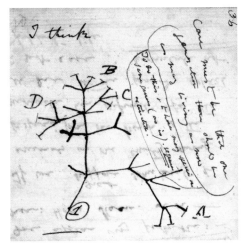

Expanding kingdoms

some blueprint for a full world designed by God to be seen as a *fait accompli*, what lay behind the natural world was a causal sequence which could be uncovered through the application of reason to evidence of present life and the fossil record. It was a paradigm shift which is hard to grasp now. The task of classification became rather different. It was no longer aiming to find out where each organism should be positioned in a hierarchical scheme, but to see how it related to present and previous organisms in both horizontal and vertical directions.

Around the same time that Darwin swapped a chain for a tree, cataloguing the natural world grew more complicated in another way. Linnaeus had settled on two kingdoms, plants and animals, leaving minerals out of his classification. But during the course of the 19th century, as microscopy improved and more microorganisms were discovered, it became increasingly difficult to maintain the idea of just two kingdoms into which all living things could be confidently sorted.

By the middle of the century, microorganisms were generally divided into

CLASSIFYING VERY SMALL THINGS

Protists are single-celled organisms that don't fit into any other classification. They comprise protozoa, single-celled algae and slime moulds. Protozoa are the earliest and most basic type of life form to have metabolized food particles. The name 'protozoa' means 'first animals' and was coined in 1818, by the German zoologist Georg August Goldfuss. He included ciliates, corals, phytozoa and medusinae.

Foraminifera are a class of aquatic amoeboid protists.

protozoa (primitive animals), protophyta (primitive plants), phytozoa (animal-like plants) and bacteria. In 1858 the palaeontologist Richard Owen (1804–92) defined plants and animals and found protozoa to share some of their features, but without the 'superadditions' of a multi-cellular organism. He referred to the 'Kingdom Protozoa' in 1860. The naturalist John Hogge objected to the term 'Protozoa' on the basis that it should really apply only to animals, and used 'Protoctista' instead. Ernst Haeckel (*see* box below) suggested the term 'Protista' for the third kingdom in 1866 and it is still used.

Haeckel considered the Protista to be primitive beings that were neither plants nor animals, but the progenitors of both. He decided that a defining feature was non-sexual reproduction and moved fungi and various other simple organisms out of the kingdoms of plants and animals and into

---

ART FORMS IN NATURE (1899–1904)

Ernst Haeckel (1834–1919) was a German naturalist and biologist. One of his greatest works was an illustrated catalogue of marine invertebrates including jellyfish, anemones (*see* picture opposite) and a type of protozoan called radiolarians. The very beautiful and intricate lithographs focus on symmetry and organization, showing how animals are similar to and different from one another. *Art Forms in Nature* (or *Kunstformen der Natur*) was extremely influential in the art world, particularly in the Art Nouveau movement, as well as in natural history.

---

protists. The definition of Protista was later refined to exclude multicellular organisms.

All was well in the three kingdoms for 60 years or so. But then, in 1925, the French marine biologist Édouard Chatton used the terms 'eukaryotic' and 'prokaryotic' for single-celled and multi-celled organisms with a nucleus (nucleated) and single-celled organisms without a nucleus (non-nucleated) (*see* box opposite). He did not make a big deal out of it, and it went largely unnoticed for several decades; he certainly didn't propose revising the tree of life to accommodate them. But in 1962, two microbiologists, Roger Stanier and C.B. van Niel, proposed dividing all organisms into prokaryotes and eukaryotes.

In 1969, plant ecologist Robert Whittaker absorbed the distinction into a five-kingdom system that set the prokaryotes apart from the four groups of eukaryotes: Monera (prokaryotes), Protista (unicellular eukaryotes), Fungi, Plantae, and Animalia. This gave the prokaryotes a kingdom to themselves, but it remained relatively unsatisfactory as the Monera is really a ragbag of all the organisms that are not eukaryotes.

Defining plants and animals

It might seem a simple thing to tell plants and animals apart, at least. Aristotle managed reasonably well. There have been glitches and anomalies – such as the vegetable lamb and the barnacle goose – but on the whole it worked until people discovered animals too small to see or difficult to observe in life.

For Aristotle, the difference lay in the type of soul that animated a being. This gives the organism its characteristics. The

BIO-SPEAK

Prokaryotic cells have no membrane-bound nucleus or cell organelles; genetic material (DNA/RNA) exists freely in the cytoplasm. Prokaryotes include bacteria.

Eukaryotic cells have organelles and a membrane-bound nucleus that contains genetic material. Eukaryotes include plants, animals and fungi.

difference between plants and animals is that animals can move and experience sense-perceptions, and plants cannot.

In the mid-19th century, Richard Owen described a plant thus: 'rooted, has neither mouth nor stomach, exhales oxygen, and has tissues composed of "cellulose" or of binary or ternary compounds.' Owen defined an animal as an organism that 'receives the nutritive matter by a mouth, inhales oxygen, and exhales carbonic acid [carbon dioxide], and develops tissues the proximate principles of which are quarternary compounds of carbon, hydrogen, oxygen and nitrogen.'

Haeckel's intricate abundance of anemones in Art Forms in Nature *clearly demonstrates similarities and differences in related organisms.*

39

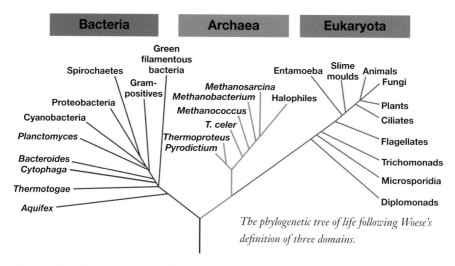

The phylogenetic tree of life following Woese's definition of three domains.

Tom Cavalier-Smith, Professor of Evolutionary Biology at the University of Oxford, defines animals as 'ancestrally phagotrophic multicells with collagenous connective tissue between two dissimilar epithelia'. Plants are organisms with 'plastids with double envelop in cytosol; starch; no phagocytosis'. Armed only with these last definitions, the average person in the street would find it hard to tell which of a giraffe and an oak tree was a plant and which an animal.

From kingdoms to domains

In the 1970s, microbiologist Carl Woese (1928–2012) was examining the genes of microbes. Woese believed he was looking at the gene sequences that should be pretty much the same in all living things, and that any substantial differences to their composition would make life unsustainable. Yet when he turned from the microbes available in his lab to some he had taken from sludge in a local lake, he found considerable and startling differences. He found the rRNA (ribosomal ribonucleic acid) patterns

in a microbe that produces methane, *Methanobacter thermoautotrophica*, very different from those of either prokaryotes or eukaryotes. Once he began to look for them, he found more microbes that shared the same signature. He named them archaebacteria. Although archaebacteria share some genetic features with bacteria, they are a very different type of organism, and are now known as archaea.

As he had found them in an extreme environment, Woese first thought they were all extremophiles, but we now know that archaea are found all over the world in any type of environment. It is significant that they can survive in extreme environments, though, as they might have been the first life forms on Earth, existing when all environments were hostile to the types of life with which we are more familiar. Their survival in such environments also has implications for astrobiology – the study of the origin of life and its (potential) existence on other planets. In 1977, Woese published his findings, redrawing the standard taxonomic tree diagram to introduce a new level, that of domains. His scheme, based not on differences in morphology (form and structure) but on phylogenetic relationships (evolutionary links based on genes), had at its base three domains: archaea, bacteria and eukaryotes.

The end of hierarchy

The chain of being is a distinctly hierarchical model of life. The lowest positions are occupied by organisms with little in the way of what we might call abilities: they can't move, can't feel and can't think (as far as we know). As we move up the chain, organisms become more versatile. From the point of view of Aristotle and most of the people who followed him over the subsequent 2,000 years, this reflects the 'fact' that some organisms are better or more fully developed than others. An elephant can do more than a worm, so it's a 'higher' animal; a worm can do more than a fungus so it's a 'more evolved' life form.

The next model, the tree of life, remains hierarchical even though its aim is to show the relationships between organisms. Organisms are fitted into the tree so that they appear on branches (and eventually on twigs) that come from dividing large branches and the trunk. The further

through the maze of branches and twigs we follow a path, the 'more evolved' the organisms become. Most people would, by default, consider a human to be a more advanced organism than an amoeba, and

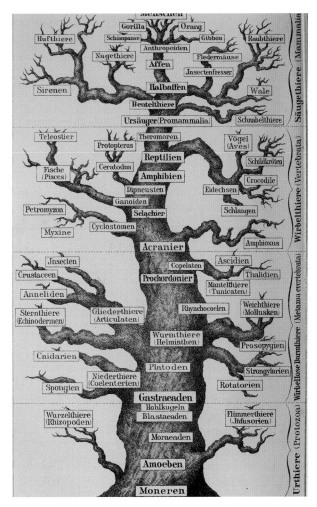

Haeckel turned Darwin's metaphor of the 'tree' of life into a vivid presentation of evolved species arranged in a tree.

even than an earlier primate. This view is almost as anthropocentric as Aristotle's graded souls, with the rational human lording it over the 'lower' animals.

The most recent model for representing the relationships between animals tries to get away from the idea of hierarchy altogether. The cladogram typically shows the end of each line of evolutionary development on the same plane as every other. This puts *Tyrannosaurus rex*, a nematode and a human as equal – and that's exactly what they are. Each is at the end (for now) of an evolutionary branch. It is not 'better' than its evolutionary ancestors, it just differs from them in ways that the cladogram shows. Cladistics was conceived by German biologist Willi Hennig in 1950 and rose in popularity after the translation of Hennig's work from German to English in 1966.

Fundamental to cladistics is the idea of the 'common ancestor'. Two species are related if they have a common ancestor, having both come about because of genetic changes to that ancestor during the course of evolution. When a species evolves a significant difference from the ancestor, the cladogram shows a branch. It then continues in two directions, and there can be further branching. It shares a lot of features with the tree of life model, but is significantly different in that it is not hierarchical – no organism is considered 'more advanced' than any other. Cladistics is directed towards showing evolutionary heritage rather than revealing how alike or dissimilar organisms are (which formed the basis of their organization into a ladder or chain).

The full cladogram of all life-forms is drawn as a circle, with all organisms on the circumference of the circle. Again, this allows no hierarchical interpretation.

Telling things apart

The definition of species, that category which ultimately determines the separation between organisms, had been problematic

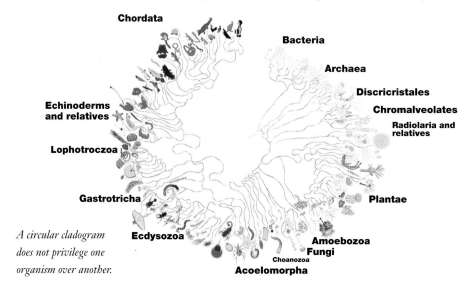

A circular cladogram does not privilege one organism over another.

Early classification systems were based on morphology – body shape – and function, so would see the similarities between bat and moth (both flying and furry). The modern classification system sees two genetically very dissimilar organisms that have evolved similar solutions (flight and fuzz) in the face of similar challenges.

for centuries, but became more acutely so with the recognition that organisms change over time. We need to distinguish not only between organisms that exist at the same time, but also between changed versions of what could be called the same organism which have existed in different times. It is a problem that is still not wholly resolved (*see* pages 194–5).

Today, taxonomy is ongoing. The fundamental principle has changed since the time of Aristotle, and even since Richard Owen's time. We no longer group organisms by their form (morphology) and responses to environmental challenges (how to move, for example). Modern taxonomy tries to establish their place in evolutionary development by also finding genetic evidence that reveals descent from a common ancestor. The earliest principles could lead us to class bats and dragonflies together, as both have wings. The second would establish that the wing of a bat has developed from the same ancestral root as the front leg of a rhino. On this basis, bats are more like rhinos than they are like dragonflies. It's a conclusion Aristotle might have found hard to accept.

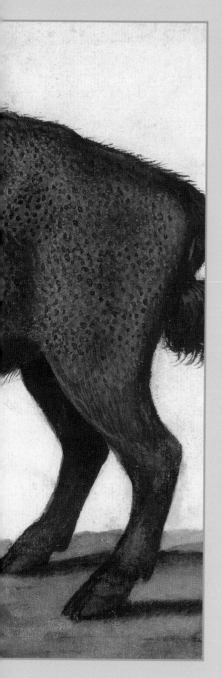

CHAPTER 2

Beastly
MACHINES

*'Study every kind of animal without distaste;
for each and all will reveal to us something
beautiful and something natural.'*
Aristotle, *Parts of Animals*, Vol. 1

**The idea that plants and animals work
in accordance with the natural laws of
physics and chemistry is common sense
to us, but was inconceivable before
the 16th century. Until then, vague
concepts such as 'vital spirits', essences
and humours were invoked to explain
the mysteries of the body.**

*Discoveries in the workings of animal bodies came at the same time
as the discovery of entirely new types of animal bodies, as European
explorers penetrated the Americas.*

Before our eyes

The origins of anatomy and physiology lay in simple observation: people identified structures before they could ascertain what they did, or observed processes but could not say how they occurred. It took a change in world view, alongside developments in technology, physics and chemistry, before the internal workings of human, animal and plant bodies could be properly revealed.

Rudimentary exploration of living creatures must have begun thousands of years ago. It can't have escaped the notice of our early ancestors, for example, that the insides of animals they cut up for food were similar to those of injured humans they had seen.

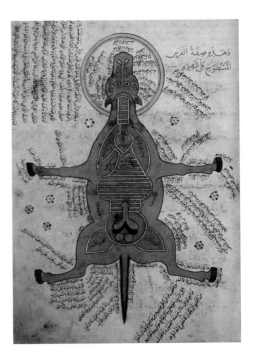

Right: The basic anatomy of a horse, depicted in a 15th-century Egyptian manuscript.

HUMOROUS PEOPLE

The dominant model for the functioning of the human body did not rely on anything discovered by dissection. It rested on the notion of four 'humours', which were thought to underlie the health of the body and the temperament of the individual. Humoral theory can be traced back at least to the school of the Greek physician Hippocrates (c.460–c.370BC) and was embraced also by the influential physician and anatomist Galen of Pergamon (*see* page 49) in the 2nd century AD. The humours were equated with bodily fluids: blood, phlegm, black bile and yellow bile. Every individual had a particular normal balance of the humours that accounted for their character. In general, well-balanced humours characterized good health and an imbalance led to poor health and illness. To restore health, the humours had to be rebalanced. The idea of balance in the body probably predates Hippocrates and is found also in traditional Indian and Chinese medicine.

Much medical treatment up to and including that in the 19th century was based on humoral theory to some degree. Dogged adherence to this theory accounts for the widespread use of bloodletting and emetics as a supposed 'cure' for any number of ills, for instance. Even so, Hippocrates' ideas represented a considerable advance on most previous thinkers, who assigned illness to the actions of the gods. He at least believed illness and health have physical causes that can be understood.

Dead but not gone

Death reduces the body to an object and makes its insides accessible to scrutiny by any who care to take a look. The Ancient Egyptians were well versed in the insides of the human body through their funerary practices, and recognized the heart, blood vessels, liver, spleen, kidneys, hypothalamus (in the brain), bladder and uterus. According to the Edwin Smith papyrus, which dates from around 1600BC, they knew that blood vessels were connected with the heart. In Tibet, the practice of sky burial, in which the corpse is cut up and left for vultures to feed on, meant that priests became well acquainted with the insides of the body. Their knowledge might have fed into Chinese and Indian medical texts.

Under the knife

A rigorous investigation of the internal organs had to wait until the routine dissection of corpses. By understanding what had gone wrong, it became possible to heal some cases of sickness and injury. But this depended on the acceptance that physical illness had physical causes. As illness was thought to be caused by a malign or disgruntled supernatural entity, knowing what happened inside the body was not much use. People relied upon prayers, rituals and sacrifices to cure disease. Human dissection was banned for centuries, because it was thought that the body was sacred or it had to remain whole for the person to enter the afterlife. In Ancient India, it was forbidden even to take a knife to the body. The 6th century BC Indian physician Susruta found a way round this: he recommended immersing a body in the river, inside a cage, for seven days. The flesh could then be peeled back without recourse to a knife and the underlying structures examined. The idea that animal bodies might work in ways similar to the human body and therefore be worthy of investigation does not seem to have occurred to anyone.

The natural philosopher Alcmaeon of Croton (5th century BC) may have been the first person in the West to carry out dissections in order to explore the

The Ancient Egyptian practice of mummification meant that the inside of the body was familiar terrain.

The Indian physician Susruta is known as the author of the oldest Indian medical text, written c.*600*BC.

There is more evidence that another Greek natural philosopher, Herophilos, carried out dissections in Alexandria in the 4th century BC. He is said to have done so in public, so that others might also learn from them, working with his colleague Erasistratus on the bodies of executed criminals. Unfortunately, the nine anatomical texts he wrote have all been lost.

Aristotle dissected many different types of animal and investigated developing chick embryos. Even though his descriptions of what he saw appear to be accurate, it is very difficult to infer function from structure when looking at a dead organism. He made some serious errors of judgment, including dismissing the brain as a rather useless, cold, wet organ that serves only to cool the blood and secrete mucus. He considered the heart to be the seat of intelligence and sensation.

workings of the body. He identified the brain as the seat of understanding, and believed that the main sensory organs were connected to it by channels. He apparently removed the eye from a dead animal and saw that the optic nerve connected the eye to the brain, but there is no record of him carrying out more sophisticated dissection.

Herophilus and Erasistratus were the first enthusiastic dissectors and vivisectors.

WHAT YOU SEE IS WHAT YOU LOOK FOR

The Chinese usurper-emperor Wang Mang called for the dissection of a prisoner in AD16 to gain medical knowledge. The butcher-surgeon who undertook the dissection found five organs (heart, liver, kidney, spleen and gall bladder), corresponding to the five planets; 12 tubes to carry air and blood, corresponding to the 12 great rivers of China; and a total of 365 components overall, corresponding to the 365 days of the year.

This 15th-century manuscript shows Galen accompanied by an assistant (with a pestle and mortar) and a scribe, in an apothecary's shop.

only two lobes); that there are holes in the septum of the heart through which blood passes from one side of the heart to the other (there are none); and that the human breastbone has seven sections (it has just three).

Men and monkeys

Classical Rome forbade human dissection or autopsy, and it was under this disadvantage that the most famous and influential physician and anatomist Galen (c.AD130–210) worked. Galen dissected macaques (a type of monkey), expecting their anatomy to be reasonably similar to human anatomy. It led him into many errors, which were then perpetuated by later anatomists who looked at the bloody, often hacked-about lumps of flesh beneath their knives and saw what Galen had led them to expect. This included a liver with five lobes, as found in a dog (while the human liver has

Back with the dead

From the Roman period until the European Renaissance, there was no progress in physiology or anatomy. Galen's lessons, based on the macaque and other animals, went unchallenged for around 1,000 years.

Perhaps surprisingly, the Christian Church was not so squeamish as Ancient Rome and India, and both dissection and autopsy were allowed and relatively common, at least from the late Middle Ages. The first anatomy textbook based on dissection of human bodies was written by Mondino de' Liuzzi (1275–1326). Indeed, dissection became a mainstream part of a physician's education, beginning in Italy around 1300. It became fully established as a vital part of medical training with the founding in 1594 of the first permanent dissection theatre at the University of Padua under anatomist and surgeon Hieronymus Fabricius.

Shock, horror – Galen wrong!

In 1539, the young Flemish anatomist Andreas Vesalius (*see* box opposite) dared to demonstrate how wrong Galen had been. Vesalius had become increasingly alarmed at the difference between what he saw when he dissected corpses and what Galen had led him to expect. When he suspected that Galen had based his work on animals, Vesalius then collected animals of many types for his own dissections. His suspicions were soon confirmed. In 1539, he publicly compared skeletons of a Barbary macaque and a human, showing the root of Galen's errors. This was not a popular move. Many people were outraged and continued to defend Galen, even against such clear evidence. A contemporary French anatomist, Jacobus

De' Liuzzi's medical treatise included a section on trephination (hammering a hole in the skull).

Slyvius, claimed that the human body had changed since Galen's time, so keen was he to protect the latter's reputation and defend the model on which he had built his own career.

Picture this

Vesalius' book *De Humani Corporis Fabrica* included detailed and accurate (if sometimes bizarre) anatomical diagrams. It was only during the 16th century that anatomy books became usefully illustrated in this way. Previous books had sometimes been illustrated, but usually either with astrological drawings or with 'wound men'. The latter showed an unfortunate individual

---

SOUL-SEARCHING

In 1533, the Catholic Church ordered an autopsy to be carried out on conjoined twins Joana and Melchiora Ballestero to discover whether they shared a soul. The autopsy found two hearts and so concluded the twins had two distinct souls, following the reasoning of Empedocles who said that the soul is harboured in the heart.

THE FABRIC OF THE BODY

Andreas Vesalius was a surgeon and anatomist – and the first modern European to insist on the importance of dissection in understanding the human body. Born in Brussels in 1514, he trained in Paris and in Padua, where he became Professor of Surgery and Anatomy in 1537 at the age of only 23. He published his most important book, *De Humani Corporis Fabrica* (*On the Fabric of the Human Body*), in 1543. It was a turning point in medical science in that it showed, through detailed drawings and descriptions, that many of the treasured beliefs of the time were wrong. He called for other doctors to carry out dissections and put their faith in what they discovered for themselves. Vesalius worked closely with talented artists to make sure his anatomical illustrations were as accurate as possible.

beset by all the many types of wound that were possible. While Vesalius's figures often adopt strange poses – half-flayed, or dangling from a gallows with their bones and organs exposed – what is depicted is at least accurate and based on the examination of real corpses.

Below left: A 'zodiac man', showing which signs of the zodiac had dominion over different parts of the body.

Below right: Vesalius showed what lay beneath the skin, laying bare muscles, ligaments and bones in a grotesque display.

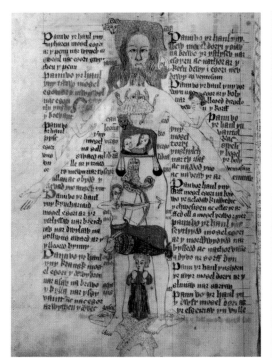
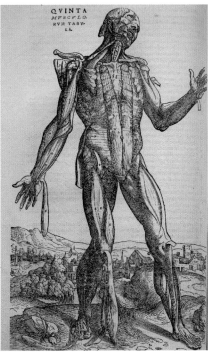

Since the 16th century, illustration has been key to the dissemination of biological information of certain types. Physiology is an obvious example. No amount of elegant or detailed description can communicate the network of blood vessels or the structure of the lungs well enough for the reader to visualize it – yet in the early days of anatomical illustration, some anatomists complained that clear diagrams would distract students from observing properly during dissections.

Before the development of printing and woodcuts, anatomical drawings had to be copied by hand by non-expert scribes. They doubtless lost accuracy at each stage, in a graphical version of Chinese whispers. Woodcuts were developed in China and brought to Europe around 1400, with the first book illustrated with woodcuts produced in 1461. Artist and print maker Michael Wolgemut (who instructed Dürer) improved the technique in 1475, but Dürer brought it to its highest level of

ART AND ANATOMY

The new artistic style of the Renaissance, using perspective and fine detail in realistic depictions of the world, made scientific illustration genuinely useful for the first time. Among the most striking practitioners were Leonardo da Vinci (1452–1519) and Albrecht Dürer (1471–1528).

Leonardo is known for his anatomical studies as well as his inventions and paintings. Dürer, an admirer of Leonardo, brought the same interests to northern Europe. Many of Leonardo's drawings remained hidden in his notebooks for centuries, so had little impact on contemporary practice or knowledge. By contrast, Dürer's anatomically accurate pictures spread far and wide.

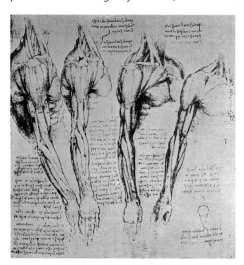

Leonardo planned a book on human anatomy and left around 200 sketches for it, but it was not published until more than 160 years after his death. Even so, his anatomical illustrations and studies show the standard of dissection and what was possible at the time. His careful observation of the musculature and anatomical structure of the horse are clear in his plans for the statue of Francesco Sforza, and his dissections also included fish, flies, moths, crocodiles, dogs, cats, birds, cows, horses, monkeys, lions, bears and embryonic chicks.

Leonardo's drawings of the musculature of the arm.

The wing of a blue roller bird, painted by Albrecht Dürer.

achievement. He was the first significant artist to sell single-leaf woodcuts as individual pieces of art. His interest in the accurate depiction of human and animal subjects led to some astonishingly beautiful paintings and woodcuts that promoted the biological sciences around Europe.

Until the invention of photography, zoological and botanical illustration were crucial to spreading knowledge. Robert Hooke's *Micrographia*, published in 1665 (*see* pages 88–90), was instrumental in bringing microscopy to public attention. The beautiful illustrations of plant material, produced with improved printing technology in the 18th century, both reflected and boosted a rising interest in botany.

The quick and the dead

It is one thing to cut up a dead body, tracing the paths of nerves and blood vessels, locating the organs and muscles, and drawing the entire internal structure of the body (as Vesalius and his artists did). It is another matter entirely to examine the workings of the living body, to see inside while it is working.

Some evidence of 'inside' is visible to everyone. Blood trickles out if you cut yourself, and a serious wound shows layers of muscle and bones and organs further within. Clearly these all do something; the question was how to discover what they did. The most unethical approach was reportedly taken by Herophilus, who was said to have used vivisection on 600 live prisoners in the 3rd century BC. Vivisection of animals has been common. Vivisection of humans has not been as uncommon it should have been. The victims were often prisoners, slaves and prisoners of war, even into the 20th century. Not surprisingly, perhaps, it's difficult to deduce much from a body that is writhing and screaming – vivisection has not been the most useful of the anatomist's tools.

In the centuries after Vesalius, biologists and physicians turned their attention to the working processes of the body – how it is nourished and grows, how the blood flows round it, how the muscles and nerves provide movement and how the sensory

'The substance of the lung is dilatable and extensible like the tinder made from a fungus. But it is spongy and if you press it, it yields to the force that compresses it, and if the force is removed, it increases again to its original size.'

Leonardo da Vinci

Descartes' diagram of the brain, spinal cord and nerves.

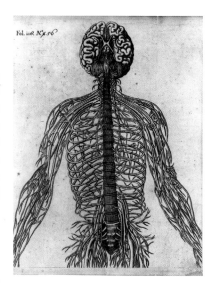

organs work. Later, with the benefit of microscopy, the fine structures of organs and tissues would be revealed. Eventually, the working of the body at the chemical level would complete the picture.

From organism to mechanism

Dissection can show the layout of the body's insides, but it does not readily show how those parts work. In the wake of the more sophisticated dissection techniques developed in the 16th century and the shift in viewpoint that came with the Enlightenment in the 18th century, a new way of thinking about the body emerged. The recognition that physical and natural laws govern processes, from the movement of the planets to the way water courses through rivers or pipes, cast human, animal and even plant bodies in a new light. Physiology and anatomy took a new turn, and it was a decidedly mechanistic one.

Tubes, pipes and valves

The 16th and 17th centuries saw great advances in mechanics and engineering which introduced, among other things, reliable clocks, firearms and even amusing automata. The last one might seem frivolous, but the French philosopher René Descartes (1596–1650) claimed to have been prompted by the sophisticated automata in the gardens of Versailles to consider that the human body might also be a system of tubes, channels, switches and valves capable of seemingly miraculous movement. The mechanical body would become the prevailing paradigm of the age. Significantly, machinery is fully explicable in terms of the laws of physics and does not rely on magic or spirit to drive it.

Increasingly in the 17th and 18th centuries, people began to see plants and animals not just as whole, mysterious organisms animated by different types of souls, but as complex biological machines. If the body is a machine, its working can be understood.

> 'Life is but a motion of Limbs, the begining whereof is in some principall part within; why may we not say, that all Automata (Engines that move themselves by springs and wheeles as doth a watch) have an artificiall life? For what is the Heart, but a Spring; and the Nerves, but so many Strings; and the Joynts, but so many Wheeles, giving motion to the whole Body, such as was intended by the Artificer?'
>
> Thomas Hobbes, *Leviathan*, 1651

This shift in perspective opened up human, animal and plant anatomy and physiology to closer scrutiny.

Weight-watcher

The idea that mathematics could be applied to bodies, as it could be applied to planets, levers and bridge-building, inspired Santorio Santorio (1561–1636), professor of theoretical medicine in Padua. Santorio followed the teaching of Hippocrates and Galen in treating his patients, but in his research he trusted first the evidence of his senses, or experience, second his reason, and third, that of previous authorities.

Santorio was one of the first practical scientists to approach the body as a mechanism governed by mathematical laws, likening it to a clock or machine. His principal achievement was in bringing mathematics and accurate measurement to the study of medicine and physiology. His exploration of what goes into and comes out of the human body represents one of the most dedicated examples of self-experimentation in the history of biology, for perseverance at the very least.

Santorio built a platform attached to a beam on which he spent a good part of his time over a period of 30 years. He had important bits of his furniture installed on the platform and sat in a chair on it to eat and drink. The construction was an elaborate weighing machine. He kept a record of his weight at all times, and also weighed all that he ate and drank, and all his urine and excrement. He found that his weight did not exactly equal his starting weight plus inputs less outputs. This suggested to him that a great part of the food he took in was not added to or expelled from his body in obvious ways but lost in 'insensible perspiration'. This was the weight lost as fluids in various ways, and the portion of his food that was expended in energy, though the second concept would have been alien to him.

Just as important as Santorio's experiment itself was his conviction that observation and reason were more important guides in biology than authority. But while Santorio established the important principles that measurement can reveal vital information and we don't use all the food we take in for growth, it could not give any insight into the actual process of digestion. That would have to wait for another development: the notion of the body as a chemical system.

Santorio lived for more than 30 years in his bizarre weighing machine.

Moving bodies

Descartes' concept of the 'mechanical body' led him to give a comprehensive but fanciful account of the nervous system. Despite having carried out many dissections, collecting the heads of various animals from a French butcher, he found what he expected to find. He described the nerves as channels, with valves, which conveyed 'animal spirits' from the brain to the muscles. This was in keeping with the 'balloonist' theory that had gas or spirits travelling through nerves – thought to be tubes – to inflate the muscles, so causing movement. It originated, like so many errors, with Galen. Descartes believed, too, that there was a thin thread running the length of these channels, and any motion of the thread could trigger the brain to make the muscles move.

The 'animal spirits' were centred in the pineal gland, which he considered to be the seat of the soul – chosen because there is only one (and we need only one seat for our soul) and because he believed (wrongly) that it is found only in humans and not in other animals. He managed to construct even thoughts, emotions and imaginations after a mechanical model using the animal spirits, but his account cannot be grounded in any empirical evidence – he constructed a scheme that he would have liked to see exist.

Descartes was a thinker, but not a rigorous experimenter. In the interaction between nerves and muscles, he had picked one of the more difficult systems to investigate, not to mention the least susceptible to mechanistic explanation. It took a more practical scientist to discover how muscles work, and that person was the Italian physiologist Giovanni Borelli (1608–79). The workings of the nerves would remain unexplained for another century.

Borelli was wedded to the mechanistic model of the body, believing that the working of the muscles 'could be solved by the almost direct application of known mechanical methods'. He carried out careful observation of muscles working in life and in death. By exposing muscles underwater and separating the fibres

---

BODIES AND SOULS

The model of the body as a mechanism presented a problem which troubled Descartes. How can the physical body, operating according to physical laws, contain and interact with the non-material soul or spirit? Yet clearly there is interaction between the two: we move our bodies following intentions formed in the mind; emotions such as misery and joy have physical manifestations; and what happens to our body affects our mental state.

The French physician and philosopher Julien Offray de La Mettrie claimed that the mind is also part of the bodily mechanism, arguing in 1745 that humans are just very complex animals. He accepted the determinism implied by his view that the mind is part of the machine: that if all our bodies and minds do is follow natural laws, then we have no free will. Descartes was unwilling to go so far, and was left with a gulf between mind (or soul) and body that was difficult to bridge.

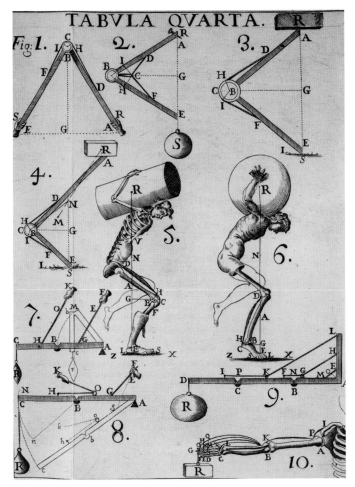

Giovanni Borelli applied geometry and mechanics to the movement of the human body.

that the swelling of contracted muscles does not represent any increase in volume, but is achieved by shortening of the fibres. Others had demonstrated that when a muscle contracts underwater there is no rise in the water level, which means there is no change of volume in the muscle. Steno worked with Dutch microscopist and entomologist Jan Swammerdam (*see* pages 92–4), and discovered that many muscles could be induced to continue working after the death of the animal, including the heart (which he showed to be muscle). This proved that no 'vital' or 'animal' spirit was involved in their action; muscles were decidedly mechanical.

longitudinally, he was able to show that no gas had inflated the muscle, for if it had it would escape in bubbles. He made many measurements as he worked out exactly how animals moved by the action of their muscles and bones, and therefore is considered the originator of biomechanics.

The Danish anatomist Nicolaus Steno (1638–86) studied muscles around the same time as Borelli, aiming to apply the mechanical principles discovered by Galileo to the movement of animals. He showed

> '*Many people talk of the animal spirits, the more subtle part of the blood, the juice of the nerves, but these are mere words signifying nothing.*'
>
> Nicolaus Steno, 1667

However, what caused the muscles to move was another matter. The Swiss anatomist Albrecht von Haller (1707–72) was interested in the form and function of various tissues and organs and is most famous for his studies of the 'irritability' of muscles and the 'sensibility' of nerves. The distinction was that the irritable organs and tissues contracted or moved when stimulated, as discovered through vivisection, while those that are sensible transmit a message to the brain when stimulated. He found that stimulating certain nerves causes corresponding muscle movements. Further understanding of how nerves and muscles operate together would have to wait for another great discovery outside the field of biology – electricity.

The anatomy of the human head, by Albrecht von Haller.

Blood and heat

The circulation of the blood is one of the few physiological systems that can be fully described mechanically.

It must have been clear since earliest times that blood is essential for life, but not exactly what it did. Galen had taught that the venous and arterial systems were entirely separate, and that the heart produced heat and the lungs served to cool the blood (and indeed the heart). In his model, the arteries dilated to draw in air and contracted to expel air and vapours through the pores in the skin. This was based on Empedocles's view of the lungs (*see* page 60–1).

Harvey – not the first

In the West, the English physician William Harvey (1578–1657) is generally credited with explaining the circulation of the blood (*see* page 59). But he was not the first to do so. That honour goes to the Arab physician Ibn al-Nafis who in 1242 suggested that blood travels from the heart to the lungs, mixes with air,

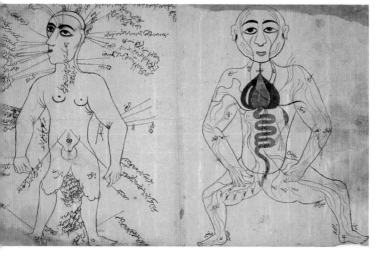

18th-century Persian drawings show blood-letting points and the venous system.

> 'Blood from the right chamber of the heart must arrive at the left chamber, but there is no direct pathway between them. The thick septum of the heart is not perforated and does not have visible pores as some people thought or invisible pores as Galen thought. The blood from the right chamber must flow through the vena arteriosa [pulmonary artery] to the lungs, spread through its substances, be mingled there with air, pass through the arteria venosa [pulmonary vein] to reach the left chamber of the heart, and there form the vital spirit.'
>
> Ibn al-Nafis, *Commentary on Anatomy in Avicenna's Canon*, 1242

and travels back to the heart, before travelling to the rest of the body. A manuscript of al-Nafis' work was discovered in Berlin in 1929, long after Harvey had been credited with explaining the circulation. It is not clear whether al-Nafis' manuscript had become known in Europe before Harvey published his own account in 1628, but it was certainly Harvey's work that had the most impact.

The Spanish physician Michael Servetus had also described blood flow to the lungs, in 1553: 'The blood is passed through the pulmonary artery to the pulmonary vein for a lengthy pass through the lungs, during which it becomes red, and gets rid of the sooty fumes by the act of exhalation.' Unfortunately, Servetus published this in a theological text, *Christianismi Restitutio*, which also contained heretical views on the Trinity and predestination. It didn't go down well. He was executed and almost all copies of his work were destroyed.

The movement of the heart

Trained at Cambridge University and the University of Padua, William Harvey practised medicine in London. From 1616, he held the Lumleian lectureship, a post with a seven-year tenure that was intended to increase knowledge of anatomy in England. He accompanied his public dissections of corpses with lectures about what they revealed, yet in his work on the circulation he realized it was important to observe and investigate the heart in action, in a living subject. He found it extremely difficult: 'I found the task so truly arduous . . . that I was almost tempted to think . . . that the movement of the heart was only to be comprehended by God.' Harvey could not, of course, cut open a person and watch their heart beating. But he could experiment more freely on other animals, and his conclusions were based on experiments and dissections of eels, other fish, snakes and chick embryos.

In 1628, he published his ideas on the circulation of the blood, *De Motu Cordis* (*On the Motion of the Heart*). His conclusion was that the heart pumps blood from the left ventricle around the body and from the right ventricle to the lungs. He worked out that the blood flows from the heart through the arteries and at some point it returns to the heart by way of the veins, having discovered that if he tied up the arteries, the heart would fill up; if he tied the veins, the heart would empty.

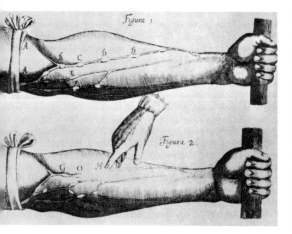

Using a ligature tied around a subject's arm, Harvey was able to investigate the flow of the blood through arteries and veins.

Harvey worked without a microscope, so could not see the system of capillaries that completes the picture. His explanation of the transition from arterial to venous circulation was entirely theoretical: that blood 'permeates the pores' of the flesh from the arteries and is gathered up from the tissues by the veins. The capillaries were finally revealed by Marcello Malpighi in 1661 (*see* pages 87–8).

By tying and then loosening a ligature around a subject's arm, Harvey was able to feel the bumps of valves in the veins. These had been described previously by the Italian anatomist Hieronymus Fabricius, who had noted them during dissections but could not account for their function. Harvey discovered that he could not force blood to flow against the direction encouraged by the valves (so back down the veins of the arm), but it was easy to make it flow up the arm. The opposite was true of the veins in the neck, so clearly the veins carried blood towards the heart.

Harvey's work was far from popular. It disagreed with Galen, and that alone was enough to cast doubt on it. It also undermined the common medical practice of bloodletting, a treatment used for almost everything, including those ailments that would seem to us least suited to the treatment, such as anaemia and blood loss. But Harvey's findings marked a turning point in our understanding of the body and how it functions – his conclusions could not be resisted for long. Beyond elucidating the operation of the heart and the circulatory system, Harvey's work was crucial in that he used experiment and observation to overturn centuries of traditional but unfounded belief.

Breath of life

Harvey did not work out what the blood was doing on its journey around the lungs, and indeed the process of gas exchange would not be discovered for a long time – during Harvey's lifetime no one was even aware that there are different gases.

The Ancient Greek philosopher Empedocles (495–430BC) asserted that all

> 'I have heard him [Harvey] say, that after his Booke of the Circulation of the Blood came out . . . he fell mightily in his Practize, and that 'twas beleeved by the vulgar that he was crack-brained; and all the Physitians were against his Opinion, and envyed him; many wrote against him.'
> John Aubrey, *Brief Lives*, 1680–93

THERE'S NO SUCH THING AS A WITCH . . .

King James I was a firm believer in witchcraft and sometimes asked his private physican, William Harvey, to investigate cases of witchcraft. On one occasion, Harvey was sent to question a woman who had been accused of being a witch. He introduced himself to her

as a wizard and said he had come to discuss the Craft, and asked if she had a familiar (a demonic helper disguised as an animal). The woman answered that she was a witch and had a toad as a familiar. She put down a saucer of milk and a toad came to drink from it. Harvey sent the woman to fetch some beer, and in the meantime he killed and dissected the animal, finding it was nothing but an ordinary toad. The woman was upset by this, but Harvey placated her by revealing that he was the King's physician sent to discover whether she was a witch and that he should have had her arrested if he believed that she was.

things breathe through tiny pores in the skin leading to bloodless channels in the flesh. When the blood – which he believed ebbed and flowed within the body – withdrew from the skin, the air rushed in through the pores. When the blood surged towards the skin, air was pushed out. It was a significant step in that he acknowledged the action of substances and structures too 'fine' to be visible, but it was entirely hypothetical and not supported by any evidence or experimentation. Yet Galen repeated it, so the notion survived for 2,000 years.

The first real understanding of respiration came with the work of the Anglo-Irish chemist Robert Boyle (1627– 91). He showed that some component of air is essential to life by demonstrating that if an animal is kept in an enclosed space and given time to use up the air itself, or if a candle has already burned out in the air, the animal will suffocate. (Oxygen was not discovered until 1772.)

Robert Hooke explored the mechanical movement of the lungs, showing that it was all that is needed to get air into the body. By opening the thorax of a dog he demonstrated that he could continue artificial respiration by alternately collapsing and expanding the lungs. Further, if he pricked small holes in the lungs to let the air out, he could continue just by pumping air into the lungs – it seemed to be the passage of air through them that was the important part of the process. This fitted well with the mechanical paradigm. How air is taken into the lungs and passed to the blood was discovered by the Italian biologist and physician Marcello Malpighi from his work on the blood capillaries in the lungs. Malpighi is now regarded as the 'father of microscopical anatomy'.

A copy of the compound microscope used by Robert Hooke c.1675.

Zombie frogs and Frankenstein's monster

While the muscles do the actual moving, it is the nerves which prompt the movement. Descartes' model of the nerves as channels carrying 'animal spirits' to the muscles had been discredited by Borelli and Steno, but it would be some time before the operation of the nerves could be correctly explained. First, the American physicist Benjamin Franklin had to harness the electricity of thunderclouds. He first tried to do so in 1752, and publicized his results in 1767.

If Luigi Galvani's mother ever warned him not to play near metal objects in a thunderstorm, he clearly paid no heed. It was the 1780s, and there was considerable interest in electricity. Galvani set up an experiment in which he used a wire to connect the hind legs of a newly killed frog to a lightning conductor. When lightning struck, the electricity travelled along the wire and the frog's legs twitched dramatically. He could produce a similar result with a a machine that generated static electricity. The experiment followed Galvani's accidental discovery while dissecting a frog that an electrostatic charge from his metal equipment caused a leg muscle to twitch. Galvani was convinced that 'animal electricity' flowed to the nerves in some electrically charged fluid to produce movement.

The Italian physicist and chemist Alessandro Volta (1745–1827) disagreed with Galvani's interpretation. He considered 'animal electricity' to sound too supernatural and unscientific, and

Galvani's experiment showed that recently dead muscle tissue can respond to external electrical stimuli. It inspired Mary Shelley to write her gothic novel Frankenstein, *some 20 years later.*

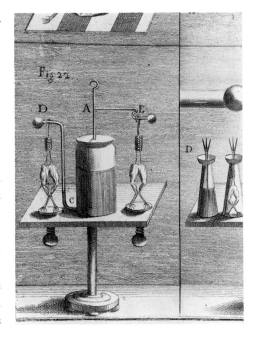

claimed that it was just ordinary, externally generated electricity acting on the muscles. Galvani went on to demonstrate that the electrical potential created by using two different metals touching the nerves of a frog leg could also make the legs twitch. The disagreement between Volta and Galvani remained civil; Volta's own experiments in the area led him to the invention of the first electric battery, the Voltaic cell, in 1800. He named the electric current produced from chemical interactions 'galvanism' in Galvani's honour. Galvani, despite having misinterpreted the situation slightly, had founded the field of bioelectricity.

How to build a body

While most of the physical processes within a body can be observed over a very short time period, nutrition and growth are rather different. Heartbeat, breathing and muscle movements take place over a fraction of a second, but digestion takes hours and growth can take years. Could they still be explained mechanically?

Anaxagoras had an oversimplified view of how food contains all the ingredients needed to build the body.

It's fairly obvious that we eat food and we grow larger. It was clear even to early thinkers that complex processes are at work, as we take in fruit, bread, meat and so on yet grow very dissimilar tissues: skin, hair, bone, blood. How can one be transformed into the other?

Sorting seeds

The Greek natural philosopher Anaxagoras (*c*.510–428BC) believed that all substances contain the 'seeds' of all other substances. So when we eat, say, pineapple, the fruit contains all that is required for it to become teeth, hair, flesh, muscle or bone. In the process of digestion, the body sorts the seeds into the right categories and sends them to the right places in the body. The seeds are assimilated by the appropriate tissues, through the attraction of like to like.

Galen saw digestion as a form of cooking or fermentation carried out in the stomach. These are chemical processes, though Galen didn't know that. It was the mechanical processes of digestion which first attracted attention.

THE CONDEMNED MEN

The Holy Roman Emperor Frederick II (1194–1250) was a highly cultured and scientifically curious monarch, but with an underdeveloped sense of ethics. One of his experiments, as recorded by the Franciscan monk Salimbene di Adam, was designed to test whether the process of digestion was aided better by rest or exercise. He invited two men to a hearty dinner, then sent one out to hunt and the other to bed to sleep. That evening, he had both disembowelled in front of him and the contents of their guts compared – 'and it was judged by the physicians in favour of him who had slept'.

Mechanics of digestion

With no understanding of chemistry, the process of digestion is quite mysterious, even if we stop short of considering how the material of food could be reconfigured into the material of bodies. The evidence from observation shows that we grind up food in our mouths, somehow extract something from it in our guts, and expel the waste as a nasty mess. Vomiting gives evidence of food further broken down in the stomach, but offers no clue as to how that happens. With the dominance of the mechanical model of the body, and no other clues, digestion was generally seen as a process of churning and grinding up food, the action begun by the teeth being continued in the gut. The maverick physician Paracelsus (1493–1541) claimed that the stomach contains an acid – 'hungry acid' – which is derived from drinking acidic spa water – and is necessary for digestion. But he had no way of proving this, and the idea had no real impact. It was not helped by the fact that few scientists took Paracelsus seriously. Although he was ahead of his time, his chemistry was mixed with a good deal of superstition and conjecture. Consequently, the mechanical model of digestion persisted virtually unchallenged until the 18th century.

The chemical body

Physiology could only make limited progress by following the mechanistic model of the body. For a complete understanding, chemistry was needed, too. Throughout the Renaissance, chemistry was essentially alchemy. It did not begin to emerge as a proper scientific discipline separate from alchemy until the work of Robert Boyle. The chemicals most important to metabolism (oxygen, carbon dioxide, hydrogen and nitrogen) were only discovered in the mid-18th century.

Paracelsus was ahead of his time in trying to investigate the body's chemistry, and was ridiculed as a consequence.

Since the early 19th century, it has become apparent that all the functions of the body are chemical at their most fundamental level. Chemistry encompasses all the metabolic cycles, operation of nerves, transport of oxygen and nutrients, homoeostatic regulation (maintaining equilibrium), the endocrine (hormone) system, growth, regeneration and far more. But to trace the discovery of all the biochemistry of animal bodies is far beyond the scope of this book. We must be satisfied with the discovery that digestion is a chemical process, as it first introduced the body's chemistry to scrutiny.

Enough to make you sick

Few of us would be prepared to go as far in the name of science as the Italian priest and biologist Lazzaro Spallanzani. An enthusiastic experimentalist, Spallanzani achieved the first recorded instance of successful artificial insemination (in a dog) and subjected many unfortunate salamanders to tail amputations during his studies of tissue regeneration. But for his experiments on digestion he was his own subject. He was not quite the first, but expanded on work first carried out by French scientist René de Réaumur (1683–1757).

Réaumur experimented with a tame buzzard, feeding it small open-ended tubes containing food and sponges. The sponges absorbed gastric juices, which he retrieved when the buzzard regurgitated the tube. Réaumur published his results in 1753, showing the effect of gastric juices in dissolving food – but only *in situ*. He found that if he extracted the juices and mixed them with food pastes, they had no effect outside the buzzard's gut. Spallanzani extended the same type of study using many types of animals, including himself. He retrieved gastric juices by vomiting, and examined the progress of digestion by swallowing small bags of food attached to threads that he could retrieve after an interval. Contrary to Réaumur's findings, Spallanzani determined that if he kept the mix of gastric juices and food at body temperature, the food was digested even outside the body. He published his results in 1777. Even so, he concluded for some reason that the gastric juices were not acidic.

Spallanzani's results were unpopular and disbelieved by many of his contemporaries. One of those who challenged them was the Scots surgeon John Hunter, though he later changed his mind when he studied the action of stomach acid dissolving the gut after death.

Spallanzani carried out several experiments on digestion in which he discovered that digestive juice contains special chemicals that are suited to particular foods.

The gut in action

A fine opportunity to observe digestion *in vivo* was presented to the American surgeon William Beaumont (1785–1853) when he treated an unfortunate fur-trapper named Alexis St Martin, who had been shot accidentally and wounded in the stomach. Under Beaumont's care, he recovered from the wound, but with a permanent fistula – a hole leading from the outside of his abdomen into his stomach. Seizing the chance for some research, Beaumont experimented on St Martin in 1825–6 and again in 1829–30. He began by dangling different types of food by a silk thread in the hole in St Martin's stomach and pulling them out again after intervals of one, two,

three, four and five hours to examine the state of digestion. He compared the rates of digestion of different types of food, and compared digestion in the gut with digestion in a test-tube using gastric juices drawn from St Martin's stomach. He found it took five times as long for a piece of boiled beef to digest in the test-tube as in the stomach, and that, unless heated, the gastric juices had no effect outside the body. He discovered that the weather and St Martin's mood could also affect the speed of digestion.

What's in food?

Beaumont's experiment showed that the body breaks down food through chemical activity. With the flourishing of chemistry in the 19th century, greater understanding of nutrition and digestion became possible. The French chemist Jean-Baptiste Boussingault (1801–87) carried out experiments in animal nutrition as part of his work in agricultural chemistry which revealed that animals don't fix nitrogen from the atmosphere but take it from their food. He also studied the proportions of some other chemicals which animals extract from food, including iron and iodine, and demonstrated that cows and other animals can form body fat from food that is rich in carbohydrate but low in fat – that is, the body can manufacture fats from carbohydrate. The principle had been established: chemical changes are effected within the body on the food taken in.

Work and fuel

It is not just food consumed that provides resources for the body. The French scientist

Antoine Lavoisier discovered that the consumption of oxygen increases when a person exercises and that a guinea pig generates heat just by existing. Lavoisier related the heat to the guinea pig's generation of carbon dioxide and compared it with a burning candle, concluding that both the heat and the carbon dioxide might be generated by the slow combustion of organic compounds in the guinea pig's tissues. His work was cut short in 1793 by a judge during the Reign of Terror in the French Revolution. Refusing Lavoisier's plea for time to finish an experiment, the judge sentenced him to death and he was guillotined the next morning.

MAGENDIE'S DOGS

'I took a dog of three years old, fat, and in good health, and put it to feed upon sugar alone. . . . It expired the 32nd day of the experiment.'

François Magendie, 1816

Magendie carried out a series of experiments in which he restricted the diet of dogs to single foodstuffs including sugar, bread, olive oil and butter. His original aim was to test whether animals could fix nitrogen from the air or must take it from their food, but it soon became apparent that dogs need more than nitrogen from food. His conclusion was that a range of varied foodstuffs is necessary for health.

Magendie was criticized for cruelty to animals during his experiments, prompting anti-vivisection legislation in Europe.

Despite Lavoisier's untimely end, the stage had been set. The body, animal or human, works on the same chemical principles as other systems. Studies in nutrition gradually revealed the variety of nutrients required by the body and that it gets energy from food. During the 19th century, the energy value of food began to be measured in calories, the unit first used to measure heat energy in chemical systems. Often, work on nutrition was carried out in animals, drawing ever-closer ties between the human body and animal bodies. When the French physiologist François Magendie (1783–1855) carried out experiments with dogs, restricting their diet to one type of food and recording the consequences, he intended the work to be directly relevant to human nutrition.

Moving inside

By the end of the 19th century it had become clear that the body combines elements of mechanics, chemistry and bioelectricity in its functioning. To understand further, it was necessary to look in detail at what goes on at the microscopic and even molecular levels within the body. Just as Harvey wasn't able to complete his picture of the circulation because he couldn't see the capillaries, so the way the nerves carry messages or the body makes use of the products of digestion couldn't be understood until cell theory was in place. The essential processes of the body take place in the tiniest units of structure, the cells, which remained hidden from human scrutiny until the 17th century and wouldn't be understood until the 19th and 20th centuries.

CHAPTER 3

What about
PLANTS?

'Plants, like algebra, have a habit of looking alike and being different, or looking different and being alike.'
Elenore Smith Bowen, cultural anthropologist, 1954

From the time of the Ancient Greeks until the late 20th century, plants were traditionally considered as inferior to, and simpler than, animals. It's easy to see why: their response to stimuli seems to be restricted to growing towards or away from something. But, as gradually became clear, this is to underestimate them. The degree to which they have been underestimated is only just emerging in the 21st century.

Blue passion flower (Passiflora caerulea) and pomegranate (Punica granatum) from Meyers Konversations-Lexikon (1897); detailed illustrations fuelled a growing popular interest in botany from the 18th century onwards.

Plants at a glance

Early farmers discovered that pollination was required for the production of fertile seeds, and that too little light, extreme heat or cold, poor-quality soil and too little or too much water could all inhibit growth. They had doubtless witnessed some plant diseases, and assault by other organisms including fungi – but this was all very much a matter of observation, without explanation or understanding. The Ancient Greeks assumed that plants took all their nourishment from the soil, but this was not based on experimentation – it must just have seemed common sense. It was not clear why plants needed light or what they might take from the soil, for instance.

Theophrastus taught and wrote on many subjects, including plants.

Plants and their uses

Theophrastus (*see* pages 18–19) spent a lot of time and effort collecting and describing plants, focusing on their differences, similarities and various uses. While he noted features innate to the plants, such as how they propagate and the length and nature of their roots, he also distinguished between them on the basis of their usefulness to humans – the provision of wood suitable for shipbuilding or charcoal, or even for making the handles of daggers, for instance.

His work includes material on topics such as the permanent and annual parts of trees and their composition; stages in the life and development of plants; plants and trees from specific areas; the effects and processes of cultivation; and the care of plants. Perhaps most importantly, he sought natural causes and explanations for the changes he observed in growing plants, rather than assuming any supernatural agency or miracle. The great Swedish botanist and zoologist Carl Linnaeus justifiably called him the 'father of botany'.

The usefulness of plants as sources of food, medicine and materials remained the focus of interest in them for many centuries. Almost all the early writing about plants is in the form of 'herbals', which give the medicinal uses of plants. Like animals, plants were sometimes found to hold divine messages, and clues to their use might occasionally be read in their forms. So the walnut, which looks something like a human brain, was considered to be good for mental wellbeing. (In fact, research in the 21st century seems to confirm that walnuts are indeed good for the brain as they contain omega-3 fatty acids – but so do many nonbrain-shaped foods, including almonds and sardines.)

Interest in the functioning and structure of plants separate from their uses was moribund from the time of the ancients until the 17th century. There is little to see inside a plant unless you have access to magnifying lenses or a microscope, so their structure remained largely unknown.

Plants as mechanisms?

Extending the mechanistic model of animal bodies to plant bodies took a leap of either faith or imagination, as there's nothing very obviously mobile or mechanical about a plant. The first person to make that leap with any conviction was Nehemiah Grew (1641–1712). He was a physician, trained in Leiden but living in London, who practised botany as much as medicine.

Grew was one of the first people to use the microscope to examine plant anatomy, publishing his findings in a series of pamphlets. He gathered

Nehemiah Grew was the first to investigate the structure of plants meticulously.

'It is noted by our Author . . . that there are those things which are little less wonderful within a Plant than within an Animal; that a Plant, like an Animal hath Organical parts, some whereof may be called its Bowels; that every Plant hath Bowels of divers kinds, containing divers liquors; that even a Plant lives partly upon Air, for the reception whereof it hath peculiar Organs. Again . . . that by all these means the ascent of the Sap, the Distribution of the Air, the Confection of several sorts of Liquors, as Lymphus, Milks, Oyls, Balsoms, with other acts of Vegetation, are all contrived and brought about in a Mechanical way.'

Review of Grew's work in *Philosophical Transactions*, 1675

them all together in 1680 as *The Anatomy of Plants*. Only 40 years earlier, the English natural philosopher Sir Kenelm Digby had denied that plants contained organs. Grew challenged this view, establishing beyond doubt that plants had distinct morphological and functional units of structure, and preparing the way for them to be properly investigated in the same manner as animal bodies were being scrutinized.

Malpighi correctly explained that galls growing on trees are the result of an insect having laid an egg in the bark, rather than of the tree spontaneously 'growing' insects.

Aside from God

Grew tried to find structures and processes in plants which were analogous to those found in animals. He looked for circulation of the sap, which could be compared with Harvey's recent discoveries relating to the circulation of the blood (*see* pages 59–61). He said growth resulted from sap carrying nutrients to parts of the plant. In all things he tried to avoid any woolly or supernatural explanation: he wanted no 'vital force' or 'sympathies' or anything comparable with the humours. Nor did he want to invoke the direct hand of God. Yet he was a believer, and wrote at length of his theological and philosophical views later in his life.

Grew was of the opinion that God had created the universe with a set of physical and natural laws which it then followed relentlessly, so no further divine intervention was needed. This is an 'intelligent design' argument of a rather hands-off type, and proposes that the universe demonstrates

the existence and skill of God in being so admirably constructed. Grew saw that much of nature was well-suited to human use (the work of the silkworm, the usefulness of iron), but realized, too, that the detailed structure of organisms is suited first of all to the needs of the organism itself. Thus his religious views were entirely in tune with his mechanistic view of nature.

Grew was not the only one looking in detail at plant structures at this time. Marcello Malpighi, who discovered the blood capillaries in animals (*see* pages 87–8), was also using his lenses and microscopes to look at plants. He was a talented draughtsman and made detailed drawings of what he saw, including the way growth occurs in plants, and the structure of the reproductive parts. Observing that if a patch of bark is removed from a tree a swelling soon appears above the bare patch, he realized that nutrients flowing down the trunk from the leaves are the source of the growth.

Plants, though, are less obviously mechanical than moving animal bodies. Most of their activity is chemical and, as such, understanding how plants work relied to some extent on discoveries in chemistry.

Water, soil, air – plant nutrition

Plants, like animals, grow from some form of seed, developing larger bodies with specialized parts – yet they don't generally eat in the way animals do. With their roots in the soil, and a fairly obvious need for water and sunlight, the issue of plant growth and nutrition was one that could be easily investigated from quite early on.

Heavy water?

Through his extended experiment recording his own weight, Santorio (*see* page 55) discovered that much of the mass of food and water taken into the human body was lost through 'invisible perspiration'. The Flemish chemist and botanist Jan Helmont (1577–1644) carried out a similar experiment with a plant and discovered that the mass of a plant increases when it takes in water. This can be deduced easily enough by observing that a pot-plant will continue to grow as long as it is watered, and the mass of the plant increases as it grows. That is indeed what Helmont did.

He dried a large quantity of soil in an oven to remove the water, then took 90kg (200lb) of it and put it into a pot. He weighed a willow sapling and planted it in the pot. For years he watered the plant with pure water, making sure the pot remained covered so that nothing else could fall into it. At the end of the experiment, he again dried the soil and weighed it, and weighed the tree. The tree had gained about 74kg (163lbs), but the amount of soil was pretty much the same as at the start. He concluded that the soil contributed little or nothing to the plant, and that wood, bark and roots had been formed from water alone.

Living on air

Helmont was not quite right, of course, because he had not taken account of what the tree takes from the air. It's rather a shame he didn't take that next step, as he was the first person to recognize that atmospheric air includes distinct gases. He discovered carbon dioxide, which he found was given off by burning charcoal, and called it 'gas sylvestre'. He recognized that it is the same as the gas produced by fermentation.

Stephen Hales (1677–1761) was the first to recognize the importance of air to a plant's wellbeing. He was an English clergyman interested in botany and the chemistry of gases (called 'pneumatic chemistry' at the time), though he began by examining the transit of water through plants. He chose a sunflower 1m (3ft 3in) tall as his experimental plant. Hales began by measuring the leaf area, root length and area of the root system. Then he measured the volume of water absorbed by the roots and lost through the leaves, and calculated the rate of transpiration (which he called 'perspiration'). He also measured the rate at which water travelled up a plant stem and the root pressure and 'leaf suction' responsible for producing 'the force of the sap' that moves it.

Hales carried out experiments with plants to measure the pressure used to move the sap.

Hales suggested that 'plants very probably draw through their leaves some part of their nourishment from the air', and pondered the possibility that they might take energy for growth from sunlight. His experiments demonstrated that plants take in air through their leaves and possibly through their trunks or stalks; Hales published his findings in 1727 in *Vegetable Staticks*.

> 'Air makes a very considerable part of the substance of vegetables.'
>
> Stephen Hales, 1727

Hales was right to suspect that air played an important part in nourishing plants, but he was not able to investigate his idea any further. That fell to others. The first was Charles Bonnet, a Swiss naturalist. He found that if a green plant is submerged in water, in sunlight, it produces bubbles of gas. He captured the gas and measured its volume – a technique still used today to measure the rate of photosynthesis. At this stage, the nature of the gas was unknown – but it was a good start.

Of mice and mint

It was already known in the 1770s that a candle in an enclosed jar would only burn for a short time before going out. The English chemist Joseph Priestley knew, too, that if he kept a mouse in a sealed jar, or introduced a mouse into the jar in which a candle had burned out, the mouse would die. He concluded that both burning candles and breathing animals somehow dirtied or 'injured' the air so that after a short time it was no longer pure enough to use for combustion or respiration.

In 1772, Priestley investigated this further. He found that if he put a mint plant and a candle into a closed glass jar, the candle would still go out after a short while, but he could relight the candle ten days later. (To relight the candle without opening the jar and introducing fresh air, he focused light onto the wick using a lens.) He also found that a mouse placed in a sealed jar with a plant lived much longer than a mouse sealed in a jar on its own. This suggested that the presence of the plant was doing something to the air, which affected the candle's ability to burn

A candle and a mouse take the same thing (oxygen) from the air; a plant can be used to restore the oxygen, making it healthy for candles and mice alike.

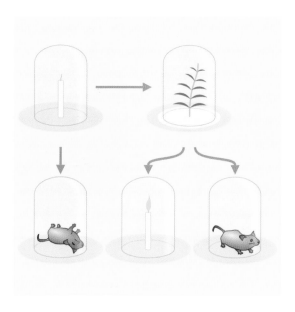

and the mouse's ability to breathe. He concluded, in the terminology of his day, that 'the diminution of the air was, in some way or other, the consequence of the air becoming overcharged with phlogiston, and that water, and growing vegetables, tend to restore this air to a state fit for respiration, by imbibing the superfluous phlogiston.' It was Antoine Lavoisier who recognized that the plant was producing oxygen, a gas he had identified in 1778 and named in 1779.

Let there be light

The Dutch physician Jan Ingenhousz repeated Priestley's experiments with plants and candles in 1778, but with an important variation: he put some of his jars in bright sunlight and kept others covered. The candles in the darkened jars fared no better than if there were no plant, but those in the sunlit jars performed as well as Priestley's original experiment. It was clear that the plants were doing something to the air, which they could do only in the presence of light. Priestley found, too, that it took just hours for the plant to return the air to a burnable or breathable state. Ingenhousz at first concluded that plants remove phlogiston from the air; but in 1796 he revised his conclusion in terms of

BURNING AWAY

Flammable materials were thought to contain a substance called 'phlogiston' that was released during the process of burning. Phlogiston was also implicated in rusting and other processes now known to be oxidation. According to the theory proposed by the German chemist Johann Becher in 1667, burning eventually stops in an enclosed space because the air can only absorb a certain amount of phlogiston before becoming saturated with it – 'phlogisticated'. The theory survived until the French scientist Antoine Lavoisier (*see* page 67) demonstrated that combustion requires a supply of oxygen. In earlier texts, 'dephlogisticated air' is oxygen and 'fixed air' is carbon dioxide.

Senebier demonstrated that the green parts of plants take in carbon dioxide.

oxygen and carbon dioxide ('carbonic acid').

Ingenhousz discovered that plants produce oxygen variably, according to the intensity of light. He found they produced carbon dioxide at night or if positioned in the shade. While only the green parts of plants produce oxygen, all parts produce carbon dioxide, and this is the same for pleasant-smelling or evil-smelling plants, and even for delicious fruit such as peaches. A person sleeping in a room stocked with fruit could, he said, be poisoned by the large volumes of gas produced. He also noted that more oxygen is emitted from the underside of leaves than from the surface.

Soon after, in 1782, the Swiss botanist Jean Senebier expanded on Ingenhousz's experiments and demonstrated that plants absorb carbon dioxide. He worked initially with aquatic plants and showed they produce oxygen only if kept in water that contains dissolved carbon dioxide. If the water is boiled (so removing the carbon dioxide), they don't produce oxygen. Senebier obtained the same result with non-aquatic plants, but held firm to his belief that the carbon dioxide came from water in the air – from humidity or from dewdrops on the leaves of the plants – rather than being freely available. He also claimed that plants release the oxygen and use the carbon to grow. (He was wrong about the oxygen, but it was a fair assumption at the time.)

He showed it was only the green parts of plants that did this – not the flowers, roots or bark. Like Ingenhousz, Senebier first formulated his ideas in terms of phlogiston, but later adopted Lavoisier's terminology. His original cycle was expressed as:

1 The plant absorbs fixed air (carbon dioxide) dissolved in water that it takes in from the ground through its roots.
2 It releases dephlogisticated air (oxygen) through its leaves.
3 The dephlogisticated air mixes with phlogiston in the air to form fixed air.
4 This falls to the soil.
5 It is dissolved in rain and groundwater and taken up again by plants.

It's a long way from the true state of affairs, but the important recognition that plants can carry out some form of chemical change in the presence of sunlight was a vital step in understanding the process now known as photosynthesis.

Senebier, to his credit, checked that sunlight was not capable of producing the same change without the presence of a green plant.

Getting chemical

The problems with Senebier's proposed cycle were clear even at the time. The first was that the small amount of carbon a plant takes from carbon dioxide can't be enough to account for its entire increase in mass as it grows. And, secondly, where does the oxygen released in transpiration come from?

The first question was tackled by a Swiss chemist, Nicolas de Saussure, who correctly outlined the process of photosynthesis in 1804. By growing plants in enclosed gas containers and measuring both the carbon dioxide absorbed and the increase in the mass of the plant, he showed that the plant increases by more than the mass of the carbon it has fixed. Something else must, then, contribute to the plant's growth. He soon demonstrated that this was water, from which the plant takes the hydrogen it needs to make hydrocarbons. By varying the amount of carbon dioxide in the air available to the plant, de Saussure found it was possible to provide so much that the plant could not absorb it and was damaged.

Further, he examined the ash of burned plants and found that they take in trace elements from the soil, and take nitrogen in more substantial quantities. The proportion of the trace elements in the soil and in the plants are not the same, indicating that uptake is selective: plants take what they need, rather than just absorbing whatever there is. This might seem rather obvious now, and with our understanding of how chemistry works it's possible to see a comparison with an animal taking what it can digest from food and excreting the waste. But a plant doesn't take in soil and excrete what it doesn't need. It's rather as if a person sat at a table and only took in the nutrients needed from the food and left the remainder untouched on the plate.

De Saussure also showed that while plants produce oxygen through transpiration, they take it in through respiration much as animals do. This had been prefigured (but not explained) by Ingenhousz's discovery that the non-green parts of plants, and plants deprived of light, produce fixed air.

Green greenery

While Senebier realized in the late 18th century that only the green parts of plants photosynthesize, it took another 50 years to discover what it was in those green parts that was doing the work. The Italian scientist Andrea Comparetti observed green granules (later named chloroplasts) in 1791, and in 1818 the French chemists Pierre Pelletier and Joseph Caventou named their green pigment chlorophyll. In 1837, the French botanist Henri Dutrochet suggested that the chloroplasts were instrumental in photosynthesis.

More links in the chain were added by three German botanists; in 1844, Hugo von Mohl described the detailed structure of chloroplasts and, in 1862, Julius von Sachs showed that the chloroplasts of a plant which has been in sunlight contain starch, whereas those of a plant which has not been in sunlight do not. This showed that the chloroplasts, through photosynthesis, are able to fix carbon into carbohydrate. This is the key to sustaining all life.

In 1848 Matthias Schleiden proposed that water molecules are split during photosynthesis, but he was not able to

PHOTOSYNTHESIS

carbon dioxide + water + light

→

glucose + oxygen

prove it. This only became possible with the discovery of isotopes (*see* box, right) and the ability to use and detect them.

The final pieces

Finally, in 1941, Samuel Ruben and Martin Kaman proved where the oxygen released by plants comes from. They supplied plants with water labelled with the heavy oxygen isotope ^{18}O and collected the oxygen the plants released. A small proportion of ^{18}O occurs naturally in the atmosphere. If plants were taking their oxygen from the atmosphere, or from carbon dioxide in the atmosphere, the emitted oxygen would have the same proportion of ^{18}O as the air around it. If the oxygen came from both the water supplied to it and from the air, it would have a proportion somewhere between the two. In fact, the released oxygen had the same proportion of ^{18}O as

Schleiden's microscopic work on the structure of plants led him to the discovery that all plants are made up of cells.

Second Lecture.

THE INTERNAL STRUCTURE OF PLANTS.

" The Great you have no power to touch,
And so attempt the Small."
FAUST.

The vignette exhibits the whole stock in trade of the scientific dealer in small wares, the microscopist ; to the right, a simple micro-scope for the preparation of small objects ; to the left, one of Amici's compound instruments ; around them, the forceps, lens, knife, razor, needles, &c.

> **ISOTOPES**
>
> An isotope is a variant of a chemical element with a different number of neutrons in the nucleus. The atomic mass of the element – shown as the superscript number in the name of the isotope – is the total number of neutrons and protons each atom has. So although all oxygen atoms have eight protons and eight electrons, ^{16}O has eight neutrons and ^{18}O has ten neutrons.

the water supplied to the plants, proving conclusively that the oxygen released in photosynthesis comes from the water the plant has taken up and not from air.

Photosynthesis is the process by which green plants fix energy from the sun into chemical energy and use it to build the carbohydrate glucose, as shown by Sachs. The energy from photons striking the leaves is stored in two chemicals in the chloroplasts: NADPH (nicotinamide adenine dinucleotide phosphate) and ATP (adenosine triphosphate). This part of the process needs light. The second part of photosynthesis can take place in the dark: the light has already supplied the energy. This involves taking the stored energy from NADPH and ATP and using it to build glucose from carbon dioxide (taken from the air) and hydrogen (left over from splitting water – the oxygen has been released to the atmosphere). The notion that photosynthesis takes place in two stages emerged in the 1930s from work originally carried out in California by Robert Emerson and William Arnold in 1932.

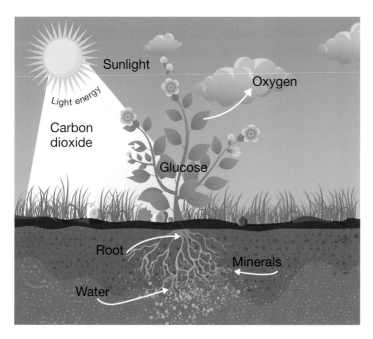

In sunlight, a plant takes water from the soil and carbon dioxide from the air, releasing oxygen and storing glucose.

In and out

Henri Dutrochet (1776–1847), who suggested that chloroplasts play a role in photosynthesis, made another vital contribution to understanding plant physiology: he discovered osmosis, the process by which liquids pass through a semi-permeable membrane (*see* box on page 80). Osmosis is vital to all organisms; in plants it enables the uptake of water through the roots. The process was first observed by the French physicist Jean-Antoine Nollet in 1748, but only fully explored in 1824 by Dutrochet, who had observed plant cells under the microscope. Dutrochet then built an osmometer to measure and demonstrate the passage of solvent across a membrane. He described how this works within plants: water enters cells in the root hairs by osmosis and moves into the xylem, the columns of fibrous cells that run all the way up the stem. He showed that water is drawn up the xylem by the pull of water evaporating from the stomata

---

IT ALL COMES FROM THIS

Photosynthesis drives most forms of life on our planet. All complex organisms ultimately rely on photosynthesis to provide oxygen for respiration and carbohydrates for nutrition. Plants, algae and cyanobacteria produce glucose using energy from the sun, carbon dioxide and water, and they use this glucose as the building blocks for cellulose. Herbivorous animals gain their nutrition from the cellulose of plant structures, and build their own bodies from it. Carnivores gain nutrition from eating herbivores (or other carnivores). All ultimately rely on the plants photosynthesizing at the bottom of the pyramid. So the discovery of the mechanism of photosynthesis was the key to understanding how life is sustained on Earth.

– holes in the leaves, which he discovered and described. Dutrochet's contribution was important in bringing together biology, physics and chemistry – in showing that the processes of plant life are subject to normal physical and chemical laws.

Growing every which way

If an animal needs food, it moves towards it; if it senses danger, it moves away from it. With a few exceptions, plants can't move like this. Their responses to stimuli generally take the form of growing in one direction or another. The ways in which plants grow directionally in response to a stimulus is called tropism. There are many forms, with plants responding to a

---

MONKS BEWARE!

Nollet was more famous as a physicist than for his work on osmosis in plants. His most notable experiment involved passing an electric current through a group of 200 monks forming a circle 1.6km (1 mile) in circumference, all connected together with iron wire to form a circuit. As each monk received an electric shock at pretty much the same time, he concluded that electricity moves very quickly!

OSMOSIS

Osmosis is the process by which a solvent (liquid or gas) moves across a semi-permeable membrane from a less concentrated solution to a more concentrated one. The membrane has holes in it large enough for the small molecules of solvent to pass through, but too small to allow through molecules of solute (whatever is dissolved in the solvent). The result is that solvent moves towards the area with a higher concentration of solute. If cells are put into water, the concentration is higher inside the cells, so water moves into the cells. The cells swell, become turgid and might eventually burst. If cells are put into a concentrated solution, water moves out of the cells into the surrounding fluid. The cells become shrivelled, and eventually the cell membrane comes away from the cell wall and the cell is plasmolyzed. If the concentration is equal inside and outside the cell, solvent passes in and out at the same rate and the cells are unaffected.

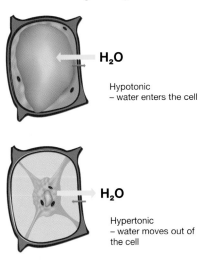

H_2O

Hypotonic
– water enters the cell

H_2O

Hypertonic
– water moves out of the cell

NASA has developed a special 'veggie' unit to experiment with growing crops in microgravity for use by astronauts on future space missions.

wide range of stimuli. The most obvious is phototropism, which causes the green parts of plants to grow towards the light. It's easy to see it happen, but much less easy to work out how the plant does it.

Early experiments revealed that a plant's roots grow down and the shoot grows upwards, even if the plant is kept in the dark. This demonstrates that the presence of light is not the only stimulus involved. In the early 19th century, the Swiss botanist Augustin de Candolle varied the water gradient so that a seed was exposed to soil that was wetter above than below, and found that the parts still grew in the usual directions – so it was not that the root was simply attracted downwards towards water. Neither does the weight of the root tip, acted on by gravity, lead to roots growing downwards, as revealed in 1806 by the British physiologist Thomas Andrew Knight. By

growing plants on rotating platforms, he showed that it is the presence of a force which determines the direction of root growth. He found that if he spun the plants quickly, their roots grew outwards (towards the edge of the circle of the revolving plate), following the direction of centrifugal force. Spun more slowly, the roots grew at an angle between the direction of gravity and the centrifugal force. Roots react positively to gravity and shoots negatively (growing away from the source of gravity). Exactly how plants sense and respond to gravity differently in different parts of their anatomy is one of the areas of botany being explored in the weightless conditions aboard the International Space Station.

Several aspects of plants' tropisms are still not understood. Roots will grow towards a pipe containing water, even though the pipe is watertight and dry on

Many food crops, including wheat and rice, are wind-pollinated.

plants, and farming techniques were developing apace. The development of the microscope made it possible, at last, to investigate the finer structures of plants and soon their sex lives were exposed to scrutiny.

the outside. They will grow away from an impenetrable barrier, such as a block of concrete, before touching it, and will grow away from the roots of a strong competitor before reaching them.

More plants

Plant reproduction has been of interest to humans since we started growing crops. Even before anyone understood what was happening in terms of plant sex, people had observed that pollination is necessary to the production of fruit and so seeds. Pictures from Ancient Egypt in 800BC show date palms being pollinated by hand with a brush – a technique that may have originated even earlier, with the Assyrians or Sumerians, and is still in use today.

Plant reproduction was not investigated in detail until the 17th and 18th centuries. Horticulturists became interested in crossing plants to gain new varieties for a buoyant market in exotic and unusual

Sex in plants

In 1676, Nehemiah Grew addressed the Royal Society in London, suggesting that stamens are the male organs of a flower and the pollen produced on them acts as 'vegetable sperm'; this was the first description of plant gendering. Experimental work by the German botanist Rudolf Camerarius (1665–1721) proved

PLANT GENDER

Most plants are hermaphrodites, having both male and female reproductive parts. A few have gendered individuals, which are either male or female, and these need both genders to be present if they are to colonize an area. In the UK, all Japanese knotweed has developed by cloning from a single individual female plant. Japanese knotweed can't set seed in the UK, but it reproduces (very successfully) asexually.

that fertilization was necessary for plants to reproduce. His work on mulberry bushes showed that if a female plant was not near any male plants, the fruits it developed bore no seeds, providing proof for Grew's theory. It seems strange that it took until the 18th century for scientists to demonstrate to their satisfaction the need for pollination, when date farmers had known it for 3,000 years.

Coloured bands or lines on flowers are often nectar guides, showing insects the path to take.

Getting together

Unlike animals, plants can't wander about looking for – and choosing – a partner. Being immobile, they have to rely on other methods of getting egg and sperm together, and they have no say in who they breed with.

The role of insects in pollination was first described in 1721 by the English horticulturalist Philip Miller, after he observed the process in tulips. The German botanist Joseph Kölreuter took this further, noticing that nectar attracts pollinating insects, and that some plants (such as grasses) rely on wind pollination. Kölreuter's work with the microscope revealed the fine structure of pollen grains and how the plant embryo develops after pollination. Another German, the naturalist Christian Sprengel, went further than Kölreuter in observing that many flowers have 'nectar guides' to draw pollinating insects to the right part of the plant. He noticed, too, that although many plants have both male and female parts they generally cross-pollinate, relying on insects to bring pollen from another plant rather than using their own pollen.

Living together

From the late 19th century, plant sciences moved into a new area, now known as ecology. It developed from 'plant geography', the study of the distribution of plants and adaptations to different environments, and grew into a discipline of its own, concerned with the interaction of plants with other organisms and the physical environment (*see* Chapter 8). It is as part of ecology that some of the surprising aspects of plant growth and behaviour have emerged.

> '*No ovules of plants could ever develop into seeds from the female style and ovary without first being prepared by the pollen from the stamens, the male sexual organs of the plant.*'
>
> Rudolf Camerarius, 1694

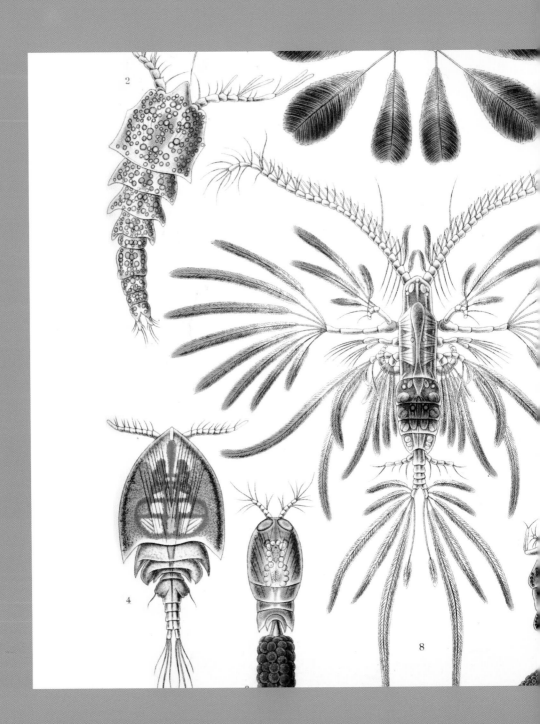

CHAPTER 4

Smaller than
SMALL

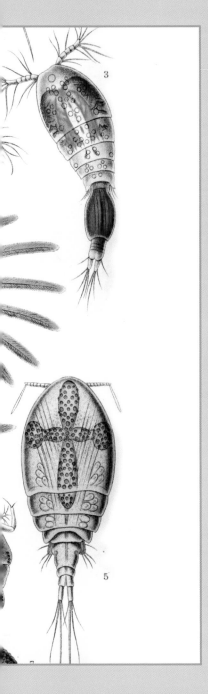

'*We must admit that there are animals a thousand times less than a grain of dust, which we can scarcely see. . . . Our imagination loses itself in this thought, it is amazed at such a strange littleness; but to what purpose should it deny it? Reason convinces us of the existence of that which we cannot conceive.*'

Nicolas Andry de Bois-Regard, 1700,
physician and writer

The late 16th century might be considered a pivotal point in the entire history of science, with the telescope and the microscope appearing within years, maybe even months, of each other. The first revealed that there are other worlds in the skies; the second uncovered the hidden realms in our own world.

There are far more microscopic organisms than there are large plants and animals, including these copepods (tiny crustaceans) drawn by Ernst Haeckel.

Imaginary small things

Even before it was possible to see very small things, some thinkers proposed their existence. Most famously, around 400BC the Ancient Greek philosopher Democritus proposed that all matter is made up of infinitesimally small particles – atoms, pretty much. The first person to propose that there might be biological entities too small to see with the naked eye that might have a biological effect was the Italian physician Girolamo Fracastoro. In 1546 he suggested that epidemic diseases are caused by something like seeds which can transmit infection with or without contact between people, even over long distances. It is not clear whether he considered the entities to be living or whether they might have been chemicals, though he refers to them as 'spores' or 'seeds':

'I call *fomites* such things as clothes, linen, etc., which although not themselves corrupt, can nevertheless foster the essential seeds of the contagion and thus cause infection.'

Into the light

The Romans were the first to discover the magnifying properties of curved lenses. In the 4th century AD, Seneca wrote of using a globe of water to magnify small script. In the 13th century, glass lenses were first used in eyeglasses to correct defective vision. The first lenses specifically intended to magnify biological specimens offered 6x or 10x magnification. This was enough to inspect small insects – and, indeed, they were first named 'fly glasses' – but not enough to see microorganisms.

The Dutch father-and-son spectacle-makers Hans and Zacharias Jansen are usually credited with making the first compound microscopes in the late 16th century, though no precise date for the first is known. Their microscopes could magnify from 3x to 10x, so were still not able to resolve microorganisms.

There has been some conjecture that Titian painted this portrait of Girolamo Fracastoro in exchange for treatment for syphilis, c.1528.

ANGELS AS THE FIRST MICROORGANISMS

Although the question of how many angels might fit on the head of a pin or the point of a needle is often cited as a preoccupation of medieval ecclesiastics, there is no evidence of their ever debating the issue. The earliest reference is in William Chillingworth's *Religion of Protestants* (1637), which mentions unnamed scholastics debating 'Whether a Million of Angels may not fit upon a needle's point?' It would appear that the notion of beings too small to see with the naked eye did not exist before the invention of the microscope, but that the first proposed microorganisms – angels – predated the discovery of actual microorganisms by at least 30 years.

Hans Jansen and his son Zacharias probably invented the microscope in the 1590s.

simple animated specks, but as beings with the complexity and even beauty of their larger relations in the animal kingdom. Puzzles in anatomy and physiology were solved or at least illuminated by looking at the microscopic structure of plants and animal bodies.

Malpighi's exploration of the body

The Italian scientist Marcello Malpighi (1628–94) was Professor of Anatomy at the universities of Bologna, Pisa and Messina. He used magnifying lenses and microscopes to investigate anatomy, looking in particular at the lungs, kidneys, spleen and liver. He spent many fruitless hours dissecting mammals before eventually hitting on the frog as his perfect subject. He

Better microscopes soon followed, and Robert Hooke, Marcello Malpighi and Antonj von Leeuwenhoek put them to world-changing use in the following century. Hooke was the first to distinguish and name cells, and Leeuwenhoek was the first to see and describe microorganisms. Malpighi looked closely at internal structures of mammalian and frog bodies.

Smallish

Early microscopes were not able to resolve the detail within most individual cells, but they allowed detailed examination of very small organisms and structures, and revealed the teeming life within common substances such as pond water and soil. Creatures such as the flea, famously drawn by Hooke, were revealed as more than

Hooke's beautifully illustrated Micrographia *brought microscopy to a wide and soon enthralled audience.*

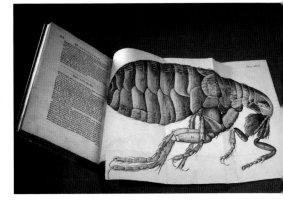

considered the frog to be the 'microscope of nature' which enabled him to see things he could not otherwise see, and remarked that, 'For the unloosing of these knots [puzzles in anatomy], I have destroyed almost the whole race of frogs.' He believed that nature first tried things out in 'imperfect' animals, such as frogs, and then after some practice used the refined technique in the 'perfect' animals (mammals and, especially, humans).

In his work on the lungs, Malpighi was able to show what Harvey could not – that the flow of blood is continuous from arteries to veins, the two joined by microscopic tubes called capillaries. His

Malpighi's drawing of a frog's lungs with (below) the alveolar surface opened out, showing the capillaries.

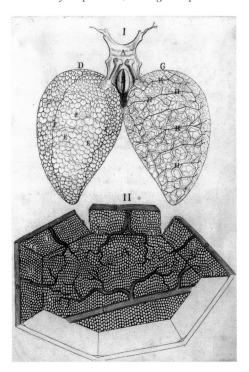

study of the lungs (in frogs and then dogs) also revealed their finer structure: wider airways resolving into ever smaller branches and finally the 'vesicles' of the alveoli. He looked, too, at the spiracles which insects use for respiration, and found that blocking them with oil caused the insect to die 'in the time it takes to say the Lord's Prayer'. Previous anatomists (even Vesalius) followed Galen in considering the lungs to be a kind of solidified bloody sponge, formed from the blood and serving to cool the organism. Although Malpighi had seen all the structures that are needed to understand how gas exchange occurs, with gas passing between the capillaries and the alveoli, he stopped just short of putting the pieces together.

Hooke and Micrographia

At about the same time as Malpighi was examining the workings of the lungs, Robert Hooke (*see* box opposite) recorded his discoveries with the microscope in beautiful and detailed illustrations collected in the volume *Micrographia: or Some Physiological Descriptions of Miniature Bodies Made by Magnifying Glasses* (1665). His astonishing drawings brought the world's attention to the realm of the tiny. But it was rather by chance that Hooke was using a microscope at all.

Originally, the English king, Charles II, had charged the architect Christopher Wren with producing a series of studies of insects viewed through the microscope. Wren made a start, but soon grew tired, disenchanted or just overburdened with other work. He passed the task on to Hooke, a young man only 26 years old, but who had a knack for drawing and using technical equipment. It turned out to be a most fortuitous act of delegation.

Micrographia presents Hooke's drawings of specimens he had examined with microscopes, and of the equipment itself. The book became the first best-selling science book. The diarist Samuel Pepys described it as 'the most ingenious book that I ever read in my life' and stayed up until 2am reading it. Not everyone saw the value of his work, though. One critic wrote that Hooke was 'a Sot, that has spent 2000 £ in Microscopes, to find out the

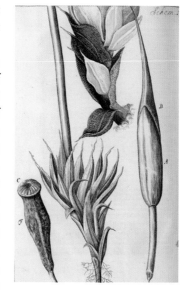

Ferns and mosses from Hooke's Micrographia.

nature of Eels in Vinegar, Mites in Cheese, and the Blue of Plums which he has subtly found out to be living creatures.' But the discovery that there are living things too small to see was revolutionary. The illustrations in *Micrographia* span a great range, from the detailed structure of everyday substances such as cloth and snow, through magnified views of small organisms such as fleas, lice and flies.

ROBERT HOOKE (1635–1703)

Robert Hooke was a home-educated child prodigy until the age of 13, when he went to Westminster School and then the University of Oxford. He excelled in technical as well as intellectual fields. As a child, he copied all the inner mechanisms of a clock, making replica parts in wood, and then put them together to make a working timepiece. He later applied his technical genius to adjusting the height, angles and illumination of the microscopes he used to gain better images than ever before. He was self-taught, too, in technical drawing, and it was through this that he revealed the microscopic world to an entranced public.

Hooke was a genius and a polymath whose true worth has rarely been recognized. Not only an accomplished biologist, he was Surveyor of the City of London after the city was destroyed by the Great Fire in 1666, an architect of many great buildings (of which few survive), the originator of Hooke's Law (the law of elasticity in a spring), the first astronomer to set out the problems in calculating the distance to a star other than the sun, and the inventor of several improvements to microscopes and clock mechanisms. He was Curator of Experiments for the Royal Society from 1661, a year after its foundation. But his irascible nature and disputes with major scientific figures such as Newton diminished his reputation. Newton blackened his name and probably destroyed the only portrait of him (none survives); he also took credit for, or hid, some of Hooke's work on gravity and light.

*'So, naturalists observe, a flea
Has smaller fleas that on him prey;
And these have smaller still to bite 'em,
And so proceed ad infinitum.'*

Jonathan Swift, 'On Poetry:
a Rhapsody', 1733

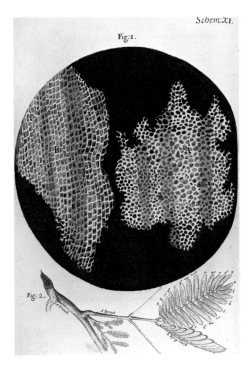

Cells revealed

Robert Hooke coined the term 'cell' for the organizational building blocks of living organisms. The first cells he described were from a sample of cork and were roughly rectangular. He adopted the term 'cell' because of their resemblance to the cells in which monks lived in a monastery. (He also referred to them as 'pores', but it was 'cells' that stuck.) He calculated that there were 1,259,712,000 (more than 1¼ billion) cells in a single cubic inch of cork.

Animalcules and others

While Malpighi and Hooke documented the microstructure of larger specimens, the Dutch lens-grinder Antonj van Leeuwenhoek (*see* box opposite) took a step further into the world of the minuscule, seeing microorganisms for the first time. His simple microscopes could produce a magnification up to 200x – far better than anything Hooke had access to – and that was enough to resolve bacteria, blood cells, sperm and the multitudes of organisms living in a drop of pond water. He made

Hooke used a razor to cut extremely thin slices of cork, enabling him to make out the individual cells.

'As to his Person he was but despicable, being very crooked, tho' I have heard from himself, and others, that he was strait till about 16 Years of Age when he first grew awry, by frequent practicing, with a Turn-Lath . . . He was always very pale and lean, and laterly nothing but Skin and Bone, with a Meagre Aspect, his Eyes grey and full, with a sharp ingenious Look whilst younger; his nose but thin, of a moderate height and length; his Mouth meanly wide, and upper lip thin; his Chin sharp, and Forehead large; his Head of a middle size. He wore his own Hair of a dark Brown colour, very long and hanging neglected over his Face uncut and lank.'

Robert Waller's account of Hooke in old age, 1705

'In structure these little animals were fashioned like a bell, and at the round opening they made such a stir, that the particles in the water thereabout were set in motion thereby . . . And though I must have seen quite 20 of these little animals on their long tails alongside one another very gently moving, with outstretched bodies and straightened-out tails; yet in an instant, as it were, they pulled their bodies and their tails together, and no sooner had they contracted their bodies and tails, than they began to stick their tails out again very leisurely, and stayed thus some time continuing their gentle motion: which sight I found mightily diverting.'

Antonj van Leeuwenhoek on the ciliate protist *Vorticella*, 1702

surprisingly accurate estimates of the size of the objects he viewed by comparing them with the size of a grain of sand.

Leeuwenhoek called the tiny moving things he saw 'animalcules', recognizing them as living things. He was not a good artist and employed an illustrator to draw for him, but wrote detailed and engaging descriptions of his specimens, many of which are easily recognizable.

Leeuwenhoek was also the first to use pigments to show up structures within otherwise transparent samples, for example, using saffron to stain muscle cells.

The ciliate protist Vorticella, *described by van Leeuwenhoek.*

His 'firsts' include:

- seeing protozoa in pond water (1674)
- discovering that yeast is an organism (1674)
- discovering red blood cells in human, fish, bird and pig blood (1675). Leeuwenhoek calculated the size of a human red blood cell as 'rather less than' the equivalent of 8.5µm – it is actually 7.7µm (1 µm or 1 micrometre is one-thousandth of a millimetre)
- examining seminal fluid from himself, and from dogs, pigs, molluscs, fish, amphibians and birds, terming the sperm he found 'animalcules' (1677)
- identifying the needle-like crystals of sodium urate that form in gout patients (1679); in 1684 he guessed that the pain of gout is caused by the crystals spiking tissue
- finding nematodes in pond water (1680)
- finding bacteria in tooth tartar and faeces, and finding parasitic protozoa in faeces (1683)
- discovering lymphatic capillaries (1683) and blood capillaries (1698)
- finding diatoms in fresh water (1702).

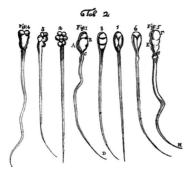

Leeuwenhoek examined his own sperm and that of various animals. He was careful to assure the Royal Society that it 'was not obtained by any sinful contrivance on my part, [but rather was] the excess which Nature provided me in conjugal relations.'

In addition, he made detailed studies of insects, discovering that the compound eye of a fly is formed of many lenses, that fleas have their own parasites, and that aphids are capable of parthenogenesis (virgin birth), with some young aphids containing fully formed offspring. He studied the stings and mouthparts of bees extensively.

Microscopes and mayflies

Leeuwenhoek's technique of using a single powerful lens was adopted by Dutch naturalist Jan Swammerdam (1637–80). Early in his career, Swammerdam had worked on anatomy, demonstrating the presence of valves in lymph vessels (now called the Swammerdam valves). He also investigated respiration, discovering the interaction of nerves and muscles (see page 62), and describing red blood cells. From the 1660s he turned his attention to dissecting insects under the microscope, in particular bees, beetles, butterflies, dragonflies, silkworms and mayflies. He was among the first scientists to study insects systematically and seriously and the first to study the stages of their

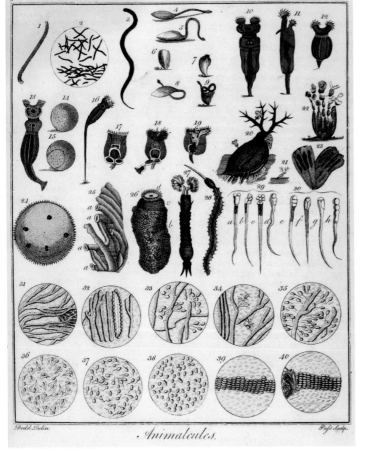

Animalcules.

A collection of various animalcules observed by Leeuwenhoek, including sperm (29 and 30).

ANTONJ VAN LEEUWENHOEK (1632–1723)

Born Thonis Philipszoon, in Delft, Holland, the boy who in adulthood came to be known as Antonj van Leeuwenhoek was the son of a basket-maker and had no formal education beyond elementary school. At 16 he began a five-year apprenticeship with a linen draper, and it was his professional work with cloth that eventually drew him to microscopy. He encountered magnifying lenses in his work in 1653 because drapers used them to count threads. These simple glasses only magnified by 3x, so Leeuwenhoek began to grind his own lenses to gain improved magnification. It is possible that he was prompted by Hooke's *Micrographia* to build his own microscopes.

Leeuwenhoek's microscopes had a single lens and a pin to hold the specimen. They were difficult to use, and had to be held very close to the eye. He is known to have made hundreds of microscopes, and even sent a batch to the Royal Society in London, though they have now been lost. He tended to make a new microscope for each sample he wanted to study, and then kept them as a permanent record. He wrote no scientific papers, but sent numerous letters describing his discoveries to the learned scientific societies of Europe.

Most unicellular organisms are microscopic, but the algae Valonia ventricosa *has a cell typically 1–4cm (¾–1½in) across*

development. He introduced a taxonomy of insects which is, in part, still used. Since Aristotle had dismissed insects as too insignificant to merit proper attention, they had been largely ignored until Swammerdam took an interest in them.

Swammerdam's diagrams are some of the most beautiful ever produced. But unfortunately he fell increasingly under the influence of the French-Flemish mystic Antoinette Bourignon, giving up science and destroying his final manuscript before dying from malaria at the age of 43.

Swammerdam developed new techniques in order to carry out his ground-breaking studies. He was the first to inject wax into specimens to keep them firm; he also dissected fragile samples underwater and used micropipettes to inflate organisms with air under the microscope. He made his own lenses and used only natural light, which

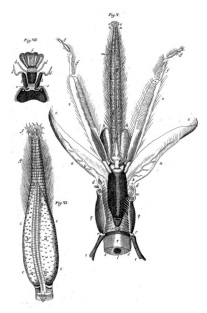

Swammerdam's detailed drawing from dissection of the mouth parts of a honey bee and wasp.

Leeuwenhoek produced the first drawings of cells. These included the nucleus in 1719, but he did not hazard a guess as to its function. There was virtually no progress in microscopic studies during the 18th century: inadequacies of the technology held back further developments and discoveries. But these problems were overcome in the 1820s.

Seeing things: colour fringes and globules

In 1824, British physicist and lensmaker Joseph Jackson Lister began work on an improved lens to view his samples more clearly. Chromatic aberrations, or 'colour fringing', frequently occurred as early lenses often failed to focus the light of different colours (or wavelengths) at the same point. Lister made the first improved microscope two years later and published his methods in 1830. Improved lenses made a considerable difference to the progress of microbiology. The early 19th century saw many reports of 'globules' forming part of the structure of biological samples, and these globules were probably circles produced by optical interference from poorly manufactured lenses.

Joseph Jackson Lister with his microscope; his son, Joseph, would introduce the use of carbolic acid as an antiseptic, making internal surgery tolerably safe for the first time.

meant that sometimes an investigation had to wait several months; most of his work took place on summer mornings. His style of single-lens microscope, like Leeuwenhoek's, was difficult to use, with the lens close to the eye and the sample close to the lens. For liquid samples, he used a thin sample tube held directly in front of the lens. For his meticulous dissections, Swammerdam used a range of tools including a saw made from a small section of watch spring, a fine penknife, feathers, glass tubes, tweezers, needles, forceps and scissors.

Cell theory

Strangely, it took until 1837 for those studying plant cells and those studying animal cells to compare notes.

One of the first people to put the new microscopes to good use was the Czech anatomist Jan Purkinje (1787–1869). He developed a knife both strong and sharp enough to cut sections from bones and teeth. He also used balsam to seal preparations, and employed methods adapted from those of the photographic pioneer Louis Daguerre to take the first photographs through a microscope. He even developed the kinesiscope, which showed pictures on a rotating drum, in order to demonstrate the operation of valves he had examined in the heart. With new lenses and new methods of preparing samples and recording what was seen, microscopists could move forward.

A cell is a cell . . .

Although many people had seen animal and plant cells, no one had yet recognized their equivalence. Animal cells are of many and varied types and it is not at all obvious that, say, a bone cell and a nerve cell are in any way analogous.

The cell nucleus was described in 1831 by the Scottish botanist Robert Brown, but then largely ignored until

Matthias Schleiden.

the German botanist Matthias Schleiden (1804–81) noticed the importance of the nucleus in the creation of new cells by division. He focused his attention on it, soon calling it the 'universal elementary organ of vegetables'. Schleiden considered cells to be 'independent, separate beings' which led a double life. A cell was at once an independent entity as well as being part of the agglomeration that is the whole plant.

Schleiden was an odd and stubborn man who had come to biology by a roundabout route: originally trained as a lawyer, he turned out to be so bad at it that he shot himself. He was no better at shooting than at law, though, and on recovering from his self-inflicted injury began to study botany and medicine. Luckily, he turned out to be better at biology.

In 1837, while dining with the physiologist Theodor Schwann, Schleiden remarked on his discovery that plant bodies are organized into cells which have independent processes and are able to reproduce themselves. Schwann realized he had seen the same in animals. It did not take

'I at once recalled having seen a similar organ in the cells of the notochord, and in the same instant I grasped the extreme importance that my discovery would have if I succeeded in showing that this nucleus plays the same role in the cells of the notochord as does the nucleus of plants in the development of plant cells.'

Theodor Schwann, writings published posthumously, 1884

Schwann carried out his work on cell differentiation in tadpoles.

of cell doctrine: 'All living things are composed of cells and cell products.' This is the first of the three basic tenets of cell theory:

- All organisms are composed of one or more cells.
- Cells are the basic unit of life.
- All cells arise only from pre-existing cells.

Schwann and Schleiden came up with the first two of these, but missed the third, proposing instead that cells form by 'free-cell formation', similar to that of crystals. This was a type of spontaneous generation (*see* pages 108–12), but at a smaller scale than had been proposed by earlier naturalists, who stated that worms and maggots generate directly from slime or mud. According to this theory, cells formed from undistinguished cytoplasm, but went on to differentiate into different cell types following 'blind laws of necessity' (a cop-out if ever there was one!) Schwann defined two categories of cell phenomena: plastic, which related to the combination of molecules that made up the cell, and

them long to check their observations in the laboratory and deduce that both plants and animals are made up of cells. It was, Schwann said, 'the most intimate connection of the two kingdoms of organic nature'.

Schwann published their findings in 1839 (without crediting Schleiden's contribution) as *Microscopic Investigations on the Accordance in the Structure and Growth of Plants and Animals*. It established the basis

FROM TISSUES TO SYSTEMS

The French anatomist Marie François Bichat (1771–1802) was a prodigious worker who is said to have undertaken 600 autopsies in a single year. Finding that breaking the body down into organs was not a sufficient level of detail to enable him to understand either its normal workings or the impact of disease, he set about 'decomposing' it into its 'intimate structures', which he named 'tissues'. He identified 21 different types of tissue in the human body. These he found organized into organs, with the organs forming the respiratory, nervous and digestive systems. As a result, some scientists considered tissue to be the final level of resolution and went on to reject emerging cell theory as either contradicting Bichat's scheme or as redundant in the light of it.

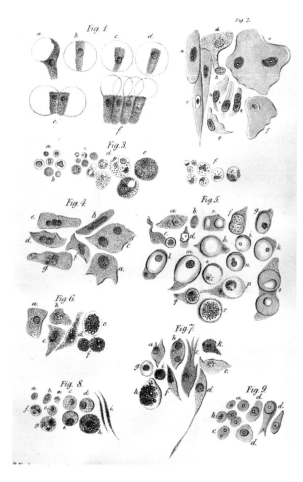

Cell theory illustrated, possibly drawn by Virchow himself.

Cells from cells

The final tenet of cell theory, that all cells come from pre-existing cells, was proposed by the German doctor Rudolf Virchow (1821–1902), who made detailed microscopic studies of cellular processes. While observing healing, he saw no evidence of cells forming from cytoplasm, but saw them created by the splitting of existing cells. The Polish-German embryologist Robert Remak had published the same findings relating to cell division in embryos in 1852; Virchow published in 1855. It's not clear whether he knew of Remak's work, but he did not cite it. Virchow stated memorably that *omnis cellula e cellula* – 'all cells from [previous] cells' – and 'there is no life but through direct succession.' This was a momentous finding, proving at last that spontaneous generation does not happen, on any level. It returned science to the position stated 1,900 years earlier by the Roman Lucretius (99–55BC) that 'Nothing from nothing ever yet was born.'

metabolic, which related to changes and processes that happen within a cell. He proposed that 'animal heat' is produced by cell metabolism.

Schwann's theory was savagely satirized by German chemist Justus von Liebig (*see* pages 189–191) in an attack that destroyed the sensitive Schwann. He gave up his studies and left the final work on cell generation for Louis Pasteur to complete in the 1860s (*see* pages 100–102).

Cell division in colour

Virchow's insight was soon confirmed by German biologist Walther Flemming (1843–1905), who developed staining techniques that enabled him to watch the actions of chromosomes in dividing cells. The name 'chromosome', coined by

German anatomist Wilhelm von Waldeyer-Hartz, means 'coloured body' and refers to the fact that they show up when stained with Flemming's aniline dyes. His conclusion, that cell nuclei reproduce, led him to state that *omnis nucleus e nucleo* – 'all nuclei come from [previous] nuclei'. Flemming didn't realize that chromosomes held genetic material. (For the role of chromosomes and the development of genetics, *see* pages 159–62.) He named the process of cell division 'mitosis'.

Eduard Strassburger (1844–1912) examined cell division in plants with the rigour that Flemming applied to animal cells. By the end of the 1880s it was clear that the nucleus splits first, with the chromosomes arranging themselves and reproducing, and then the cell divides – a process found to be the same in plants and animals, in embryonic and adult cells. Each cell division produces an exact copy of the original cell, with the full set of chromosomes and cell structures.

MODERN CELL THEORY

The current basic statements of cell theory are:

1 All known living things are made up of cells.

2 The cell is the structural and functional unit of all living things.

3 All cells come from pre-existing cells by division (spontaneous generation does not occur).

4 Cells contain hereditary information that is passed from cell to cell during cell division.

5 All cells are basically the same in chemical composition.

6 All energy flow (metabolism and biochemistry) of life occurs within cells.

In sickness and in health

Virchow was adamant that the study of cells would be the key to disease. He argued against the previous concept of disease rooted in humoral theory and said that disease must be located in the body's cells. The state, behaviour and processes of the cells were, he claimed, what differed between healthy and diseased states.

The ability to see microorganisms at last answered another question related to disease that had defeated medical science for thousands of years. How are diseases transmitted?

Early models of sickness had relied on Hippocrates' model of the four humours, or on the idea of 'bad air' or 'miasmas'. In the Hippocratic model, the body must have a good balance of the four fluids or humours (blood, bile, yellow bile and phlegm) in order to be healthy; if the humours are out of balance, sickness results and health can only be restored by restoring their equilibrium. This led to the often harmful use of excessive bleeding and purging to treat all kinds of ailments. But it also failed to explain contagion, since if disease sprang from an imbalance within the individual body, why would others exposed to a sick person also fall ill?

As we have seen, Fracastoro suggested in 1546 that disease might be caused by 'seeds' or 'spores', though it's not clear whether he believed these to be biological entities rather than chemical particles. French physician

DEATH BY SAUSAGE?

Rudolf Virchow was opposed to what he considered excessive spending on the military. This annoyed the German chancellor Otto von Bismarck so much that he challenged Virchow to a duel. As the challenged party, Virchow was entitled to choose the weapons. He chose two pork sausages: a cooked one for himself and an uncooked one, containing larvae of the parasitic roundworm *Trichinella*, for Bismarck. Bismarck declined, deciding that it was too risky.

Fungal parasites such as this on a mosquito can quickly overwhelm and kill their host.

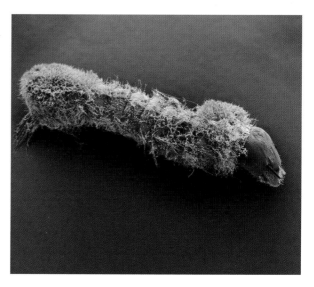

Nicolas Andry de Bois-Regard (1658–1742) was the first to propose microorganisms as a cause of disease. Developing Leeuwenhoek's work, but with specifically medical aims, Andry theorized that diseases including smallpox were the result of 'worms' in the human body. His book, written in 1700 and translated into English in 1701 as *An Account of the Breeding of Worms in Human Bodies*, includes 'spermatic worms' (spermatozoa) as well as those he thought might cause disease. The English botanist Richard Bradley took up the baton in 1720, suggesting that plague and 'all pestilential distempers' were caused by 'poisonous insects' too small to see without the aid of a microscope.

This was all so much conjecture, though, until in the early 19th century the Italian entomologist Agostino Bassi proved that a microorganism could cause disease. In this case, the disease was muscardine, which affected silkworms, and the microorganism,

which he called a 'vegetable parasite', was a fungus later named after him, *Beauveria bassiana*. Bassi spent 25 years investigating muscardine, eventually publishing his findings in 1835. The disease plagued first the Italian and then the French silk industry, which was all but abandoned by 1849. Bassi recommended keeping a good space between rows of feeding caterpillars, destroying affected caterpillars, using disinfectants and keeping farms clean. It was the same disease, and the same threat to the silk industry, that was a spur to the French microbiologist Louis Pasteur's work in the 1860s.

Pasteur and microorganisms

Louis Pasteur is usually credited with being the first to demonstrate that the germ theory of disease is correct. Even though Bassi (a great influence on Pasteur) had already shown that a microorganism, in the form of a fungus, caused disease in silkworms, it was Pasteur who went on to demonstrate many types of microbial action.

Silkworms featured again. The French silk industry was under pressure from two diseases, which Pasteur identified as being caused by microbes. One was viral, but the other was caused by a microsporidium (fungal parasite). He went on to show that fermentation is caused by microorganisms (yeasts), that food spoils because of microbial action (bacteria) and that diseases are caused by germs (in the form of bacteria, viruses and fungi). He developed the process now called pasteurization to treat foods (particularly milk) to prevent spoiling: the product is heated to a temperature high enough to kill any bacteria present, while air is excluded to prevent the entry of more bacteria.

The humble yeast

Yeast has been a hugely important microorganism throughout human history. Its action of breaking down sugars to produce alcohol, gases and/or acids is used in the production of bread and alcoholic drinks. Yet it was not clear until 1846 that yeasts are microorganisms. In 1840, the German chemist Justus von Liebig noticed that yeasts produce fermentation, but was unaware that they were alive. As a chemist, he was interested in the molecular processes going on. At this stage, the idea that a biological process and a chemical process

Louis Pasteur made huge advances in microbiology in the 1860s.

could be the same thing – that biological processes are effected through chemistry – was contentious.

In 1846 Friedrich Lüdersdorff, another German chemist, reported that yeasts are microorganisms which convert sugar into alcohol. He demonstrated that the action of the living organism is needed – he found that if he destroyed the yeast cells, their action stopped. In 1857 Pasteur concluded that living yeast was required, but did not offer a mechanism by which fermentation took place. Yet soon after, in 1860, the French chemist Marcellin Berthelot found if he treated killed yeast appropriately, the extract would bring about fermentation. He concluded that the living yeast produces a chemical which causes this to happen:

'I think that this plant acts on sugar not because of physiological activity, but simply by means of the ferments that it has the property of secreting . . . In short . . . it is seen clearly that the living being is not the ferment; but it gives rise to it. Also, once the soluble ferments are produced, they act independently of any further vital act; this activity shows no necessary correlation with any physiological phenomenon.'

The point emerging from the work of Lüdersdorff, Pasteur and others was that yeast is a living organism, though tiny, and it is capable of effecting chemical changes. The debate about the need for the living organism to be present continued to rage during the rest of the 19th century.

Germ theory takes over

While Bassi and Pasteur made huge strides in promoting germ theory, it was Robert Koch (1843–1910) who finally quashed alternative models of disease and laid the foundations of modern microbiology. Working first on anthrax, he isolated the bacterium that causes the disease, *Bacillus anthracis*. He fixed samples on slides, used stains to show up the different cells in the sample, and found the causative agent.

This was the first time that a specific microorganism had been conclusively linked with a specific disease, as Koch demonstrated the virulence of the bacterium experimentally. He began by taking a sample of *Bacillus anthraci* from a sheep that had died of anthrax. He extracted the bacteria and injected them into a healthy mouse; the mouse developed anthrax. He repeated the experiment again and again, finally publishing his conclusion – that *Bacillus anthracis* causes anthrax – in 1876.

Koch refined his methods of growing cultures and identified the bacteria that cause tuberculosis and cholera. He improved on Pasteur's technique of growing bacteria in broth, adding gelatine and agar to create a solid medium, and growing his cultures in flat glass dishes designed by his assistant, Julius Petri.

Budding yeast reproduces by a daughter cell forming as a bud on the side of the parent cell. The parent cell's nucleus divides, part migrating into the daughter cell, which then splits off.

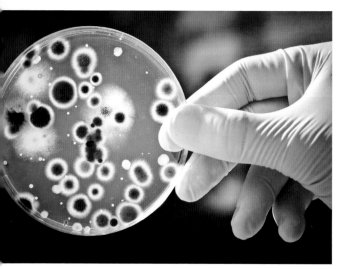

Julius Petri's design of dish is still used today to culture bacteria.

Even smaller

It seemed that Pasteur and Koch had solved the problem of microscopic organisms causing diseases. But the story was about to get more complicated.

From 1879, the German agricultural chemist Adolf Mayer worked on mosaic disease, which affects tobacco plants. He found it could be transmitted from one plant to another if he made a solution of macerated leaves from an infected plant, passed the solution through a paper filter and applied it to a healthy plant. He concluded that it was a bacterial infection.

A few years later, in 1884, French microbiologist Charles Chamberland, who worked with Pasteur, developed the Chamberland-Pasteur filter to remove all bacteria and other cells then known from liquids. It was a porcelain filter with pore sizes of 0.1–1 micron. When the Russian botanist Dmitri Ivanovski worked on mosaic disease in 1892, he filtered his solution through a new porcelain filter. His solution remained infectious, even though any bacteria should have been removed. He concluded that there was some kind of toxin in the solution that was able to pass through the filter.

It was the Dutch scientist Martinus Beijerinck who put the pieces together correctly in 1898. He, too, found a solution of tobacco mosaic virus remained virulent after passing through a porcelain filter, and decided that there must be another kind of contagious agent besides

---

KOCH'S POSTULATES

Koch set out four conditions which must be met if an organism is to be identified as the cause of a disease. These still stand:

1. The organism must always be present, in every case of the disease.

2. The organism must be isolated from a host containing the disease and grown in pure culture.

3. Samples of the organism taken from pure culture must cause the same disease when inoculated into a healthy, susceptible animal in the laboratory.

4. The organism must be isolated from the inoculated animal and must be identified as the same original organism first isolated from the originally diseased host.

IS A VIRUS ALIVE?

A virus does not have even one cell, which means it is not an organism. Is it alive? That is debatable. It can replicate itself and has genetic material, but not having an actual cell seems to disqualify it. Some biologists consider viruses to be on the border between living and non-living entities.

Viruses replicate only while inside the cell of a host organism. Outside a host cell, a virus exists as a viral particle with a two- or three-part structure – a strand of DNA or RNA, encased in a protein coat called the capsid, and sometimes with an envelope of lipids around the outside. Viruses affect all forms of life, from the largest vertebrates to tiny, single-celled organisms. Viruses that infect bacteria are called phages, or bacteriophages.

A computer model showing the structure of a tobacco mosaic virus particle.

out, rather than a contagious fluid. Viruses are typically about one-hundredth the size of an average bacterium, so remained invisible until the invention of the electron microscope in the 20th century.

Cells are complicated

During the course of the 19th and 20th centuries, improved microscopy showed more complexity in cells. The first organelles (structures within the cell) were found, though their function was not always immediately understood. In fact, the organelles are vital. Some carry out the biochemical processes of life, including metabolizing foods (from whatever source) to release and store energy.

bacteria. He called it a *contagium vivum fluidum*, or contagious living fluid, convinced that the agent itself was liquid and contained no particles. Beijerinck discovered, too, that he could keep the fluid for years and it remained infectious.

Although Beijerinck is usually credited with the discovery of viruses, Friedrich Loeffler and Paul Frosch should share the accolade. They were first to discover foot and mouth disease, a virus that affected animals, also in 1898. They proposed it was a very tiny particle too small to be filtered

The mighty mitochondrion

The most important organelle is the mitochondrion, which is often called the 'powerhouse' of the cell because it is responsible for producing most of the ATP (adenosine triphosphate) which cells use as a

SEEING CELLS

One of the most surprising and still puzzling organelles discovered in any cell is the ocelloid developed by a type of plankton called a warnowiid. The ocelloid appears to be an eye – but one so sophisticated that scientists at first thought it was the eye of something the warnowiid had eaten. It has a cornea, lens and retinal body. Its likely function as the warnowiid's own eye was only described in 2015 by a team of zoologists at the University of British Columbia.

store of chemical energy. The mitochondrion was probably first observed in the 1840s, but only established as an organelle in 1894 by the German pathologist Richard Altmann, who called mitochondria 'bioblasts'.

The American anatomist Benjamin F. Kingsbury first linked the mitochondrion to cell respiration in 1912, but it took until 1925 to work out the sequence involved. The complexity of the process of cell respiration continued to unfold during the 20th century.

In 1952, high-resolution photographs taken through a microscope showed the structure of the mitochondria and that they vary from cell to cell. Some cells contain no mitochondria (red blood cells are an example), whereas others contain thousands. Until 2016, biologists believed that all organisms have mitochondria. Then a eukaryote was discovered with no trace of mitochondria; it is *Monocercomonoides*, a microorganism found in low-oxygen environments.

An electron microscope can be used to view viruses and tiny structures within cells.

Seeing smaller, seeing more

There is always a limit to the resolution possible with an optical microscope, and therefore to the size of object that can be viewed clearly – about 0.2µm (or 200nm). The limit is caused by diffraction (spreading) of light waves and is related to the wavelength of visible light, so there seemed to be no way of getting round it. But it is possible to see smaller objects using a beam with a smaller wavelength than visible light. A beam of electrons provides a solution – the faster the beam travels, the shorter its wavelength, which can be far smaller than the wavelength of light, so it gives a higher resolution image before diffraction becomes a problem. A modern electron microscope can produce a resolution of 0.2nm – a thousand times smaller than the resolution of the best optical microscopes.

The development of the electron microscope began in 1931 with the work of the German electrical engineers Ernst Ruska (1906–88) and Max Knoll (1897–1969). They were the first people to magnify an electron image, though that was still some way from a successful electron microscope. Ruska built a prototype electron microscope in 1933 with a resolution of 50nm, but it was not practically useful. The earliest viable electron microscopes gave no better resolution than a good optical microscope and had the added disadvantage that the beam made the sample so hot that any non-metal specimens were charred. The next step was to treat samples with osmium and cut them into very thin slices to prevent charring. By 1938, there was a useful electron microscope, but World

The head of a damselfly, photographed through a scanning electron microscope.

War II held up further development. By the 1960s, resolution of 1nm was achieved. Modern electron microscopes can provide x2,000,000 magnification, which is enough to see single large molecules.

A living specimen cannot be examined with an electron microscope, so it was thought impossible to watch processes as they happen at a molecular level.

And smaller again

Using new techniques, which garnered a Nobel Prize in Chemistry in 2014, physicists Stefan Hell and William Moerner, together with Eric Betzig, found a way of using a light microscope as a sort of flashlight to illuminate individual fluorescent molecules. The samples need not be dead. Some powerful electron microscopes can now even be used to observe individual atoms, and the bonds between them. It seems that we can now see the smallest operators in biological systems – all we have to do is work out the meaning of what we see.

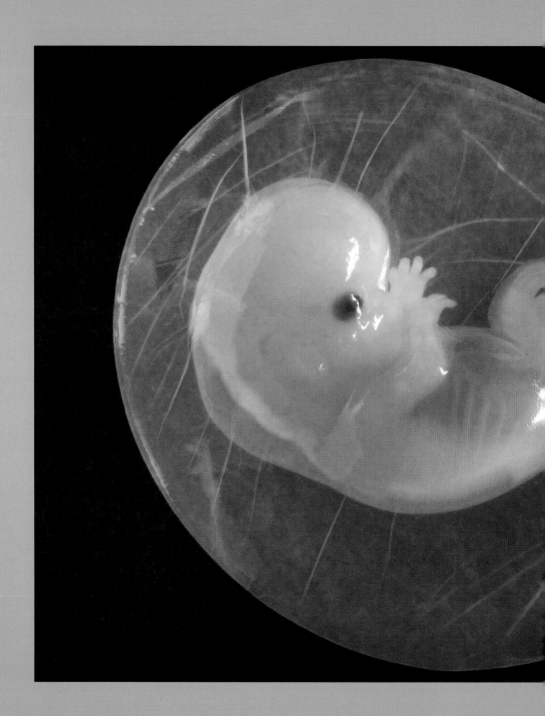

CHAPTER 5

New Life
FROM OLD

'Nothing will come of nothing'
William Shakespeare, *King Lear*,
Act I, Scene 1

The gross mechanics of reproduction in humans and large animals have always been apparent; but the nitty gritty – the biological reality of how new life is made – remained a mystery for many years.

A human embryo at seven weeks of gestation already looks recognizably like the organism it will eventually grow into.

Fertile dust

If a fly lands on a carcass or a piece of rotting food, there is an interval of a few days before maggots appear. So it's not altogether surprising that for a long time people made no connection between the fly and the maggot – which does, after all, look very different. Instead, people assumed that maggots were simply produced out of the rotting flesh. Similarly, fleas were thought to generate from dust, and worms from mud. Some people even believed that bread and cheese wrapped in rags would bring about the generation of mice, since if the parcels were unwrapped later, mice could often be found inside them. This process, known as 'spontaneous generation', was accepted without question for thousands of years.

If an animal swallows the egg of a tapeworm, the tapeworm hatches and lives in the animal's gut; it is not generated by the animal's gut.

Something from nothing after all

Aristotle, so very careful to examine the processes of embryonic development, subscribed to the popular view of his time that some animals can emerge spontaneously from non-living matter through the action of *pneuma*, or 'vital heat'. The type of creature produced depended on the conditions in which it was growing. Oysters, he thought, grew in slime, while scallops and clams grew in sand and limpets and barnacles grew in dips in a rock.

Although he believed that some animals could generate from the action of heat on certain types of inert matter, some could, according to Aristotle and others, be produced inside other animals. This was the only way to account for clearly distinct creatures such as tapeworms, which grow inside animal bodies. With no

'Now there is one property that animals are found to have in common with plants. For some plants are generated from the seed of plants, whilst other plants are self-generated through the formation of some elemental principle similar to a seed; and of these latter plants some derive their nutriment from the ground, whilst others grow inside other plants, as is mentioned, by the way, in my treatise on Botany. So with animals, some spring from parent animals according to their kind, whilst others grow spontaneously and not from kindred stock; and of these instances of spontaneous generation some come from putrefying earth or vegetable matter, as is the case with a number of insects, while others are spontaneously generated in the inside of animals out of the secretions of their several organs.'

Aristotle, *On the Generation of Animals*, Book V, Part 1, 4th century BC

understanding of the parasites' life cycles, the most logical explanation was that they were generated by the host animal.

A slow change

Challenges to spontaneous generation were slow to come. For medieval Christians the Bible seemed to support the notion, with lines in Genesis such as 'Let the waters bring forth abundantly the moving creature that hath life' (Genesis 1:20) and Adam being created from clay. Shakespeare referred to the spontaneous generation of snakes and even crocodiles from mud in *Antony and Cleopatra*: 'Your Serpent of Egypt, is bred now of your mud by the operation of your Sun: so is your Crocodile' (This was just clumsy, since snake and crocodile eggs are big enough to see.)

When the idea was criticized, it was often not for scientific reasons. Jan Swammerdam, the 17th-century Dutch naturalist, rejected spontaneous generation not on the grounds of biology or logic, but because he thought it a blasphemous idea.

Spontaneously generating doubts

The first real challenge to the notion of spontaneous generation came from the Italian biologist Francesco Redi. In 1668, in the new spirit of experimentation, Redi set out to prove that maggots did not generate directly from rotting meat but were the consequence of flies laying their eggs on it.

He took three pieces of meat and put each into a jar. One jar he left open to the air, another he covered with netting, and

EELS – A TRICKY CASE

Aristotle claimed that eels did not have sex and had no orifice from which spawn or eggs could be released. This would make any kind of normal generation quite tricky, so he concluded that they came from earthworms. Pliny the Elder, 400 years later, argued that instead they reproduced by budding, the adults scraping the new elvers off themselves by rubbing against rocks. Around the start of the 2nd century, Athenaeus wrote of eels exuding slime which, settling on mud, generated new

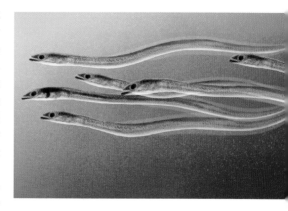

eels. He also contested Aristotle's view that anchovies come from roe, arguing that they are born of sea-foam. It's understandable that the origins of eels remained obscure since eels travel far from Europe to spawn and so eel copulation or eggs were never witnessed. Indeed, many aspects of the eel's life cycle remain obscure even today.

A RECIPE FOR SCORPIONS

'When water from the purest spring is placed in a flask steeped in leavening fumes, it putrefies, engendering maggots. The fumes which rise from the bottom of a swamp produce frogs, ants, leeches, and vegetation . . . Carve an indentation in a brick, fill it with crushed basil, and cover the brick with another, so that the indentation is completely sealed. Expose the two bricks to sunlight, and you will find that within a few days, fumes from the basil, acting as a leavening agent, will have transformed the vegetable matter into veritable scorpions . . .

'If a soiled shirt is placed in the opening of a vessel containing grains of wheat, the reaction of the leaven in the shirt with fumes from the wheat will, after approximately twenty-one days, transform the wheat into mice.'

Jan Baptist van Helmont, 1671

the last he sealed fully. Unsurprisingly, for us, the open jar was soon crawling with maggots. In the jar covered with netting, the maggots hatched on the underside of the net, the flies having landed on it and laid eggs through the holes. The sealed jar had no maggots. This proved fairly conclusively that maggots don't form from meat, but hatch from flies' eggs. Redi's work was greeted with enthusiasm by many scientists; only three years later the naturalist John Ray wrote to the Royal Society saying that Redi had 'gone a good way in proving' that organisms are not spontaneously generated.

This wasn't sufficient to convince, however. While creatures that could clearly be seen with the naked eye (flies and mice, for example) did not generate spontaneously, there was no way of proving that the tiny microbes the microscopists had found did or did not. Georges-Louis Leclerc, Comte de Buffon, was pretty sure they did. In 1777, he described the way in which the organic molecules in a decaying corpse are released and wander around until captured to become part of another substantial body. In the interim, they are available for making microorganisms:

'During the interval in which the organic molecules roam freely within the matter of dead, decomposed bodies . . . these organic molecules, always active, rework the putrefied substance, appropriating coarser particles, reuniting them, and fashioning a multitude of small organized bodies. Of these, a few, like earthworms and mushrooms, resemble relatively large animals

Francesco Redi proved conclusively that flies are not generated spontaneously from meat.

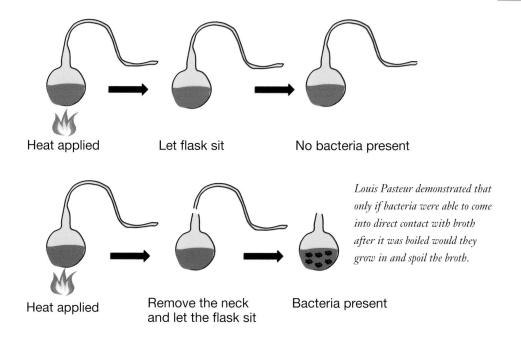

Heat applied   Let flask sit   No bacteria present

Heat applied   Remove the neck and let the flask sit   Bacteria present

Louis Pasteur demonstrated that only if bacteria were able to come into direct contact with broth after it was boiled would they grow in and spoil the broth.

or vegetables, while the others, almost infinite in number, are visible only under a microscope. All such bodies come about only by spontaneous generation.'

But evidence was growing in support of the opposing argument. Around 1729, the Italian botanist Pier Antonio Micheli discovered that if he took spores from a fungus and placed them on slices of melon, the same type of fungus would soon grow on the melon. He concluded that fungi are not produced by spontaneous generation. In 1768, the Italian biologist Lazzaro Spallanzani boiled broth in sealed flasks from which the air had been expelled. He found, as he expected, that the broth did not spoil for as long as the flask remained sealed. Opponents took issue, though, with the removal of the air. Perhaps the microorganisms in the flasks had died from lack of air?

Pasteur: taking the spontaneity out of generation

It was French microbiologist Louis Pasteur (*see* pages 100–102) who would say, resoundingly, that not even microorganisms generate spontaneously from matter. In a lecture to the Sorbonne in 1864, Pasteur contested an experiment carried out by Félix-Archimède Pouchet, director of the Natural History Museum at Rouen, which he claimed showed that microorganisms are not and could not be carried in the air. Pouchet, he showed, had not effectively excluded airborne contaminants from this experiment, so had introduced the bacteria he was claiming were spontaneously generated.

In his own experiment, Pasteur boiled broth in two flasks with curved swan necks. Afterwards, he broke the neck off one flask, exposing the cooling broth to the

> 'Never will the doctrine of spontaneous generation recover from the mortal blow of this simple experiment. There is no known circumstance in which it can be confirmed that microscopic beings came into the world without germs, without parents similar to themselves.'
>
> Louis Pasteur, 1864

air. The other he left intact. The broth in the broken flask soon began to spoil, and Pasteur showed the presence of microbes in it. The broth in the intact flask did not spoil. Although air could enter the neck, the microbes could not move against gravity to climb up the other side of the bend – they simply fell to the bottom of the curve and stayed there. It was a pretty conclusive experiment, but Pouchet and other enthusiasts continued to argue for the spontaneous generation of microbes. It took a few more decades before the last of the spontaneous generationists either changed their views or died.

Starting at the beginning

If nothing will come of nothing, as King Lear says, where does new life come from?

Hippocrates (460–370BC), considered the father of Western medicine, believed that male and female semen mix together in the female's body after coitus and the embryo develops from this. Aristotle, around 50 years later, gave the female parent a much less interesting role. His treatise *On the Generation of Animals* is the first comprehensive theory of reproduction and embryonic development. Aristotle maintained that reproduction in animals relies on the male sperm to provide the essence and nature of the new organism, and the nutritive female menstrual blood to provide the material from which it is made. This was essentially what Aeschylus had said in 458BC: that the male was the parent, and the female a 'nurse for the young life sown within her'.

Faulty males

It might be assumed that if the female contributes nothing to the nature of the organism, all offspring would be male and the system would fall at the first hurdle, with the species dying out after one generation. But Aristotle explains that pregnancy will only result in a male which is an exact copy of the father if gestation goes perfectly according to plan. Disturbances in the pattern can lead to imperfect offspring, including females, males that look like the mother, females that look like the father, or throwbacks who look like a distant ancestor.

Pasteur's flask had a 'swan neck' so that bacteria couldn't pass from the outside to the liquid within, but were trapped in the neck.

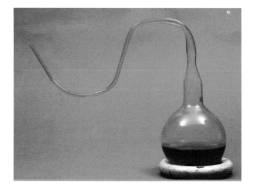

A medieval depiction of Hippocrates reading.

Aristotle's model for generation was based on his theory that there are four 'causes' of existence: the final cause, formal cause, material cause, and efficient cause. The final cause is an entity's ultimate purpose; the formal cause is its essence or existence; the material cause is what it's made from; and the efficient cause is what brings it into existence. In looking at reproduction, the first two are quite close together: the final and formal causes of reproduction are to produce a new organism. The material cause is what produces the organism (the mother's menstrual blood) and the efficient cause is what causes the new organism to be like it is, which is dictated by the father's contribution. It seemed to make sense that the embryo was built from or nourished by menstrual blood, since menstruation stops during pregnancy and it was thought that the blood must go somewhere else.

Four ways

Aristotle divided animal reproduction into four types. Animals might be viviparous (give birth to live young), oviparous (egg-laying) with hard-shelled eggs, oviparous with shell-less eggs or ovoviviparous, such as fish that hatch from eggs within the mother's body. He gave accounts of coition and embryonic development in the different types. But he also claimed that some types of creature could appear spontaneously, with spontaneous generation as an alternative to sexual reproduction in those species.

All in order

Many studies in embryology have been carried out on chicken's eggs. These are easy

Sperm attach to a human egg cell – only one will penetrate and fertilize the egg.

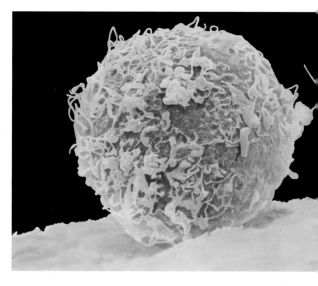

to obtain, quick to develop (taking around three weeks to incubate to term) and, for most biologists, readily disposable without too many qualms. Aristotle examined developing chicken eggs and decided that a chick embryo takes matter from the yolk to form its body. He also concluded that the parts of an embryo develop in a set sequence, with the heart coming first, then the other internal organs, and then the external features. He suggested that the order of development relates to the importance of each part in the developed organism. The idea that the body develops slowly, forming organized matter from homogenous matter, is called 'epigenesis'. The modern account of epigenesis explains how cells in the early embryo begin to differentiate and give rise to different tissues and organs as the embryo develops, following in a set sequence. Early adherents of epigenesis were often wrong about the order of development, and some thought some organs emerged from others, but the basic principle was sound.

It was possible to observe the development of non-human embryos (usually chicks), and occasionally anatomists had the chance to see a partly formed human, but there was no possibility of further insight into the start of the process of generation until the invention of the microscope.

Body and soul

In the case of humans, one particularly influential assertion made by Aristotle was the point at which the rational soul arrives in the foetus. This was, he said, at 40 days' gestation for the male foetus and 80 days' gestation for the female foetus. Christian philosophers, particularly Thomas Aquinas, took this latter as defining the point at which a foetus becomes an ensouled human. (The Catholic Church maintains that ensoulment happens at the moment of conception.)

Little progress was made in terms of practical embryology before the 16th century. The writers of the Middle Ages who concerned themselves with the unborn were most interested in the theological aspects, such as how and when the foetus was ensouled and how the character and form of offspring might reflect the spiritual health of the parents. The German philosopher Hildegard of Bingen (1098–1180) gave little

A FIRST EXPERIMENT IN EMBRYOLOGY

Perhaps the first recorded instance of embryology research comes from the Hippocratic texts, c.460BC. The experimenter is instructed to take 20 or more hens' eggs, give them to hens to sit on, and from the second day onwards take out an egg and break it to observe the development of the embryo. 'You will find everything as I say in so far as a bird can resemble a man. He who has not made these observations before will be amazed to find an umbilicus in a bird's egg. But these things are so.'

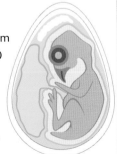

The ensoulment of a human foetus is shown taking place in this manuscript of Hildegard of Bingen's Liber Scivias. *The soul travels along a golden tube from heaven to the woman's womb.*

solace to parents unfortunate enough to have a baby born with a deformity, for it was deemed their own fault: 'The thing born therefrom is deformed, for parents who have sinned against [God] return to [God] crucified in their children.'

The first person to make detailed anatomical drawings of the developing human embryo was Leonardo da Vinci, in 1510 and 1512. His illustrations show the placenta of a cow, but the human embryo is well observed.

Deer and deer eggs

William Harvey is best known for his work on the circulation of the blood (*see* pages 58–61), but he also studied embryology. He spent many years examining chick embryos and dissected pregnant deer from the King's hunting grounds, finally publishing *Exercitationes de Generatione Animalium* (*Essays on the Generation of Animals*) in 1651. It was not well received. Harvey was thorough in his attempt to provide experimental proof of Aristotle's account, but his findings didn't support Aristotle's claim that the embryo begins to develop immediately after mating and grows from the mass of menstrual blood enlivened by semen. Harvey dissected female deer at increasing intervals after mating, and found there was no visible evidence of an embryo until at least six or seven weeks had passed, nor was there any mix of blood and semen. (He was working without a microscope, so this meant nothing visible to the naked eye.) His observations did confirm that development of the embryo proceeded by epigenesis, with the organs developing in sequence and the embryo looking increasingly like a fawn as time passed. He came to the conclusion that a vitalizing

BY-PASSING THE FEMALE

Paracelsus, a rogue itinerate medic of the 16th century, produced a recipe for making a human infant without recourse to a woman. He recommended taking semen and allowing it to putrefy for 40 days, possibly (his meaning is not entirely clear) while heated in a horse-dung-fueled incubator. Then it should be fed with human blood for 40 weeks. There is no record of whether he ever tried this method.

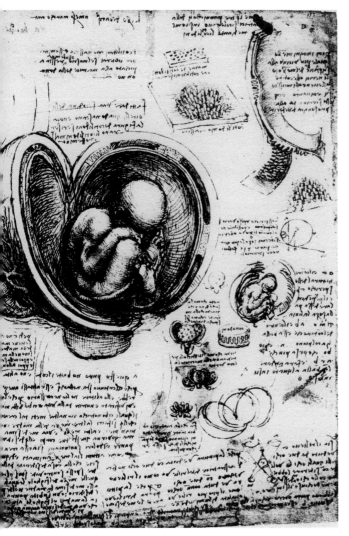

A drawing by Leonardo da Vinci of a human foetus in the womb.

century BC, claimed that:

'Everything in the embryo is formed simultaneously. All the limbs separate themselves at the same time and grow, none comes before or after another, but those which are naturally bigger appear before the smaller, without being formed earlier.'

This is preformationist theory. Aristotle's account generally found more favour, and held sway until the 17th century. Then, in spite of Harvey's findings, preformationism grew in popularity. Ironically it was the drive to make biology explicable and wrest it away from spiritual notions that led to the popularity of a model that was counter to all experimental evidence.

Galen (see page 49), on the other hand, gave a remarkably modern account of epigenesis:

'Genesis is not a simple activity of Nature, but is compounded of alteration and of shaping. That is to say, in order that bone, nerve, veins and all other tissues may come into existence, the underlying substance from which the animal springs must be altered; and in order that the substance so altered may acquire its appropriate shape and position, its cavities, outgrowths, and attachments, and so forth, it has to undergo a shaping or formative process.'

'seminal aura' invigorated the material which would become the new organism. This was unmechanical, the opposite of the model that Harvey's work on circulation had promoted.

Preformed or growing?

Although Aristotle had correctly inferred that the embryo develops in stages, the Hippocratic writings, dating from the 5th

> *'In the seed are enclosed all the parts of the body of the man that shall be formed. The infant that is borne in his mother's womb has the roots of the beard and hair that he shall wear one day. In this little mass likewise are all the lineaments of the body and that which Posterity shall discover in him.'*
>
> Seneca (3BC–AD65), *Questiones Naturales*, Book III, Chapter 29

The tension between the two accounts continued for centuries. The existence of the controversy was openly acknowledged: in ruling on the ecclesiastical law against abortion, the bishops of the Quinisext Council, held in Byzantium in 692, remarked: 'We pay no attention to the subtle distinction as to whether the foetus is formed or unformed.'

Infinitesimal parts

The drive in the 17th century to see bodies as mechanical was a challenge for theories of generation. How could the creation of an entire organism be handled mechanically, with the bits growing in the right places? How could generation be framed in a way that didn't require some 'vital spark' of the type that Descartes wished to deny? As it happened, the emergence of differential calculus and the idea of the infinitely divisible suggested a solution: preformation could supply what was needed.

If things can be divided into incrementally smaller parts, again and again, as calculus indicated, there was no reason why generation might not start off with a very tiny version of the final offspring that slowly grew bigger once its development was initiated. French philosopher Nicolas Malebranche (1638–1715) brought together the lessons of calculus and preformationism

to suggest that in fact all the creatures that will ever live were created by God in a single act of Creation and that future generations lie nested within one another, rather like Russian dolls, each unfolding in turn.

Even rudimentary observation with a microscope should have rapidly overturned the theory of preformation, yet its revival persisted for over two centuries. As late

> *'The germ of an organized body, the seed of a plant, or the embrio of an animal, in its first discoverable state, is now found to be the future plant or animal in miniature, containing every thing essential to it when full grown, only requiring to have the several organs enlarged, and the interstices filled with extraneous nutritious matter. When the external form undergoes the greatest change, as from an aquatic insect to a flying gnat, a caterpillar to a chrysalis, a crysalis [sic] to a butterfly, or a tadpole to a frog, there is nothing new in the organization; all the parts of the gnat, the butterfly, and the frog, having really existed, though not appearing to the common observer in the forms in which they are first seen. In like manner, every thing essential to the oak is found in the acorn.'*
>
> Joseph Priestley, 1803

Malebranche's theory had each generation already present within its parent, like Russian dolls.

as 1803, the chemist Joseph Priestley (*see* pages 74–5) was still repeating essentially preformationist explanations for the growth of the embryo.

Mothers and fathers

If each new organism is preformed within a parent it must be lurking within either the father or the mother. Clearly, one parent must have a much more important role than the other if one is providing the whole organism. Preformationists fell into two camps: the ovists believed the preformed progeny were latent in the mother, and the spermists believed they were present in the father. (The terms 'ovist' and 'spermist' were not used at the time.)

'Ex ovo omnia'

William Harvey (*see* pages 115–116) coined the phrase '*ex ovo omnia*' ('everything from the egg') in 1651, but this was not quite the great insight it now seems. Harvey's teacher, Fabricius, had suggested that most animals come from eggs in one form or another. Following his own embryological studies, Harvey concluded that all come from an egg – but by that he did not mean an egg cell as we know it. Instead, he was considering the egg to be 'a certain corporeal something

CHEESY SEEDS

Albertus Magnus (*c.*1200–80) was a Dominican friar and bishop who wrote on a wide range of subjects, and dissected various fish and chick embryos in his investigations of embryology. He believed that the 'seeds' of the woman coagulated when exposed to semen, rather like cheese coagulating when rennet is added. This coagulated globule developed into the embryo through contact with nourishing menstrual blood (which Aristotle had claimed provides the matter for the growing embryo).

Albertus Magnus was responsible for bringing the works of Aristotle to wider attention in the 13th century.

having life *in potentia*' and seems to have viewed the early-stage embryo (as soon as it was visible to the naked eye) to be the 'egg'. That really makes his statement incontestable as it means, essentially, that everything grows from its earliest detectable form. It did, though, rule out spontaneous generation, which was a considerable leap of faith for the time.

In 1666–7, the Dutch biologists Jan Swammerdam (*see* pages 92–4) and Jan van Horne worked on insect reproduction and on the mammalian uterus. Using Swammerdam's technique of injecting wax into soft structures to preserve their shape, the pair carried out a series of dissections, coming to the conclusion that the ovaries produce eggs and not, as previously thought, some kind of female seminal fluid. (Harvey had been unable to find anything – either eggs or 'semen' – in the ovaries.) The eggs, they decided, are moved by peristalsis (the type of muscle movement that forces food through the gut) to the womb. If the egg encounters sperm in the womb, it develops into a baby, the sperm providing the essential living soul.

Swammerdam's work on insect and frog metamorphosis seemed to support preformation. He found the legs of the tadpole ready to erupt before they are visible, and his dissection of a butterfly chrysalis showed the entire butterfly folded up inside. He concluded that it had probably been there from the start, hiding within the caterpillar and presumably the egg.

The microscopist Malpighi (*see* pages 87–8), who observed capillaries and alveoli, trained his lenses on the developing chick embryo and revealed details of earlier stages

A caterpillar turns into a butterfly inside a chrysalis, but there is no fully formed butterfly hiding inside the caterpillar.

of development than had previously been described. Although he did not explicitly state support for preformation, he claimed to have seen a preformed chick in an unincubated egg. Since he lived in Italy, where an egg lying in the hot sunshine might possibly start to develop, this was not really conclusive proof of preformation, but it did skew the debate in the direction of ovist preformation at the time.

The Swiss naturalist Charles Bonnet (1720–93) was the first to demonstrate experimentally that aphids can give birth to live young by parthenogenesis, which he took to be conclusive proof of preformation. This had long been reported but never confirmed by experiment. His studies revealed that females hatched in the summer gave birth, but those hatched late in the year bred with males and laid eggs. He managed to produce a line of nine generations by parthenogenesis and concluded from his success that every female contained within her the 'germs' of all her future descendants. The germ would grow into progeny when properly nourished, and one method of nourishment commonly used among 'higher' organisms was by sperm.

When Bonnet extended the theory to other organisms, he hit a few problems. He was very devout, and when his experiments with hydra and polyps in 1741 revealed that new organisms could be generated from limbs removed from a parent, he was troubled by the implication that this meant the soul (of the hydra or polyp!) was not, then, indivisible and unique. His solution was to suggest that the germs in these animals are scattered around the body (hence a new animal could grow from an arm or other fragment), but the germs do not in fact contain the essence of a unique individual. Instead, they hold a kind of species blueprint and the characteristics of the individual are formed by external factors such as diet and the environment of the mother's body. Bonnet, like Malebranche, wanted to believe that all generations of all organisms were created by God in one fell swoop at the beginning of the world.

Championing the sperm

Sperm were first reported by Stephen Hamm and Anton von Leeuwenhoek in 1677 (*see* page 91). Aided no doubt by the room for imagination afforded by their poor-quality microscopes, some 'saw' confirmation of performationism, with tiny humans folded into the head of the 'animalcules' in semen. The 1695 drawing of a sperm by Dutch physicist Nicolaas Hartsoeker (opposite) shows how easy it was to do.

Leeuwenhoek, on the other hand, saw the 'little animals of the sperm' in semen and decided they probably migrated to the ovaries where they fed and grew to look like eggs. He became a champion of spermist preformation even though he witnessed parthenogenesis in aphids in 1677. You would imagine that would have swayed Leeuwenhoek in favour of

---

A FINE EXPERIMENT

Bonnet demonstrated parthenogenesis with a careful experiment that at the time was considered exceptional in its design and careful execution. It had been proposed by the French scientist René Antoine de Réaumur (1683–1757), but he had not been able to make it work.

Bonnet trapped a single aphid and isolated it in a jar with a leaf. He then observed his single aphid from 4 or 5am until 10 or 11pm every day, recording its activities (though it's hard to imagine that it had many activities to choose from). Within the month, the first of his 'little prisoner's' progeny was born; a further 94 would follow over 21 days. This single experiment on a single aphid was sufficient for Bonnet to be lauded by the Academy of Sciences in Paris; he became its youngest corresponding member. In subsequent experiments, he raised successive generations of aphids from a single virgin mother.

ovist preformation. But he was so unwilling to concede the vital role of the male that he contrived an explanation: the aphid is really a spermatic animalcule – and so male – and that is why it contains its own progeny. Nice try.

Putting the pieces together

Spallanzani (*see* pages 65–6), who was in close correspondence with Bonnet, was another ovist. His discovery that frog eggs grow larger while still inside the mother was, he considered, proof that the infant frog pre-existed in the unfertilized egg – otherwise, why would the egg grow larger? He was a dedicated experimentalist, though, and proceeded to show that sperm were essential to the growth of the embryo.

Linnaeus had stated that fertilization never takes place outside the body, but Spallanzani proved this was not the case through his experiments on frogs, starting in 1771. He killed frogs during coitus and compared the development of eggs that had encountered sperm with those dissected from within the frog's body. He made special, tight taffeta pants for frogs, then allowed them to try to mate; he found the eggs from these couplings did not develop. And he artificially inseminated frogs' eggs with the sperm he collected from the pants. (Washing the sperm out of frog pants has to be one of the weirder jobs a biologist has done!) His conclusion was that semen is necessary for fertilization, and that fertilization can be external and artificially arranged – he also carried out artificial insemination with toads, silkworms

Hartsoeker's imaginative depiction of a sperm shows a tiny proto-human curled in the head.

and a dog. But he was not able, despite much experimentation, to determine whether the fertilizing agent was the liquid of the semen, the sperm (which he considered to be a parasite) or some 'aura' like magnetism, as Harvey had claimed. He judged the question to be beyond human understanding.

On the other side of the epigenesist/preformation fence, the German physiologist Caspar Wolff (1733–94) also admitted that some things were beyond human capability to know. In his case, it was the mechanism by which fertilization and embryonic development come about. He was a proponent of observation, and gave detailed accounts of what he saw as he studied developing plant and animal embryos. His conclusion was that development does occur and can be charted, with undifferentiated material finally becoming differentiated and resolved into structures. He argued in 1759 that the embryo develops from layers, though he was not aware of the role played by cells. His microscopes were not good enough for him to be more precise, and he tended to see what he was looking for: he began with

'One day what you have discovered [artificial insemination] may be applied to the human species, to ends we little think of.'
Charles Bonnet to Spallanzani, 1781

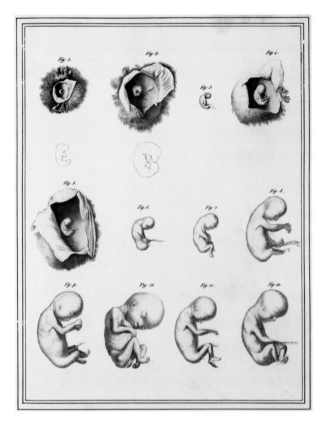

Early foetal development stages up to four months' gestation, from a French anatomy atlas by physician and surgeon Jules Germain Cloquet, 1825.

changes in the blastoderm during early development. He discerned and described in detail the three layers crucial to the development of the organism: the outer (or serous) membrane, now called the ectoderm; the middle (or vascular) membrane (or mesoderm); and the inner layers – mucous membrane (or endoderm). Pander worked out that the blood vessels developed in the middle layer.

Eggs abound

Despite Harvey's comment *ex ovo omnia* ('everything from the egg') in 1651, it was not until the 19th century that the mammalian egg was actually found and viewed through the microscope. This is quite surprising, given that the egg is the largest single cell in the mammalian body.

The German biologist Karl Ernst von Baer was the first to extract a mammalian egg, when in 1826 he discovered eggs in the ovaries of a dog – in fact, in the dog of the professor of physiology, Karl Burdach. The Swiss anatomist Albrecht von Roelliker recognized sperm and eggs as cells in 1840. But it was another century before the human egg cell was extracted and observed, by American anatomist Edgar Allen, in 1928.

an anti-preformationist agenda and found evidence of change and development. He stressed that scientists could not say why epigenetic development happened, only that it does.

From blob to being

In fact, the embryo does develop in layers. The blastoderm – the outermost layer of cells in an embryo – was discovered by the German embryologist Heinz Christian Pander in 1817 and initially named 'Pander's nucleus'. Using hundreds of fertilized chicken's eggs maintained at the optimum temperature by a team of technicians working round the clock, Pander studied

> 'With the formation of the blastoderm, the whole development of the chicken in the egg is founded and, from that time on, it progresses and concerns only this blastoderm; for every remarkable event that can happen afterwards must be considered nothing else but a metamorphosis of this membrane.'
>
> Christian Pander, 1817

Baer also introduced the term 'spermatazoa' for the sperm visible in semen, replacing the the word 'animalcules'. Unaware of the function of sperm in reproduction, he, like Spallanzani and others before him, thought they were parasites. Baer went on to study the development of the embryo, coming up with four laws of embryology.

Karl von Baer formulated the laws of embryology in 1828.

The key insight, that the sperm fertilizes the egg, came in 1875 with the work of German zoologist Oscar Hertwig, on sea urchins (*Echinoidea*). He discovered that the two cells fuse into a single cell and that it is from this that the embryo develops. He noted, too, that only one sperm cell is needed to fertilize the egg, despite the very large number of sperm produced.

Although Hertwig is always credited with the discovery of the one-egg/one-sperm recipe for fertilization, there are earlier accounts of fertilization that seem to have gone unnoticed. One of these was by the French naturalist August Alphonse Derbès who, in 1847, observed and described the fertilization of sea urchin eggs:

'[T]he spermatozoids advance[d] progressively towards the eggs. Some of them were soon encircled by a compact mass of moving corpuscles; others, farther away, only found themselves in contact with a very small number; in both cases, I saw the signs of fertilization.

'The first apparent effect of this union is the almost immediate appearance of a perfectly transparent envelope that encircles the yolk at a certain distance, which is manifested by the appearance of a circular line. I saw this envelope manifest when in contact with a

BAER'S LAWS OF EMBRYOLOGY

1. General characteristics of the group to which an embryo belongs develop before special characteristics.

2. General structural relations are likewise formed before the most specific appear.

3. The form of any given embryo does not converge upon other definite forms, but separates itself from them.

4. The embryo of a higher animal form never resembles the adult of another animal form, such as one less evolved, but only its embryo.

very small number of spermatozoids (three or four, sometimes even one only).'

Ontogeny recapitulates phylogeny

The issue of how an embryo develops was not entirely answered by the discrediting of preformationist theory. The question remained of how the organism grows from a single fertilized egg cell to the complex shape it becomes when born or hatched. Watching the development of an embryo reveals that it goes through some strange-looking stages which bear little relation to the appearance of the final organism.

German biologist Ernst Haeckel (1834–1919) was one of several scientists who proposed, in the wake of Darwin's work, that an embryo, as it grows, recapitulates the evolution of its species. So the embryo of a bird or mammal looks at one stage like a fish, because the animal's evolution can be traced back to fish. Haeckel called his theory the 'biogenetic law', and published it in

1866, just seven years after the publication of Darwin's *On the Origin of Species* (*see* pages 150–153), a book he actively championed in Germany. According to Haeckel, embryological studies could reveal the evolutionary history of an organism. He thought it possible also to see at which point species diverged in evolution by examining the progressive development of

Sea urchins are easy to find on a dive or in a rock pool, and are invaluable to biologists.

SEA URCHIN AS MODEL ORGANISM

Sea urchins are echinoderms (which means 'spiny skin'); they are small, spiny animals found in all the oceans of the world. They have been used as a model organism for the study of embryology since the late 19th century because they offer several advantages to the biologist. First, they are readily available from the sea – it's always best to pick an organism that's easy to get hold of. They are small, and reasonably robust. More importantly, their embryos are transparent, so their development is easily observed with a microscope. Sea urchins release up to 20 million eggs into the water, which are fertilized by free-swimming sperm. The fertilization of the egg and development of the embryo are straightforward, the organization of the animal is simple, and its development is rapid. In addition, it's quite easy to manipulate and disrupt the development of the sea urchin, and to observe the effects of disruption, so they are ideal for embryological and genetic studies.

their embryos. Haeckel was not the only person to suggest a recapitulation theory.

Baer's fourth law of embryology was aimed at recapitulation theory, which had already been proposed in a pre-evolutionary version by Johann Meckel in 1808. Meckel had said that embryos go through stages which resemble the adult form of less complex organisms lower on the *scala natura*.

Splitting and growing

Another solution to the question of how the embryo grows was suggested by German experimental embryologist Wilhelm Roux, who founded 'developmental mechanics'. His mosaic theory proposed that when the first cell divides, the daughter cells produced are not identical but are primed to become different parts of the organism. In 1888, he claimed to have demonstrated this by destroying one half of a two-cell frog blastomere and watching the remaining cell develop into half an embryo. He concluded that development is mechanistic, rather like building a machine.

Mosaic theory was soon disproved. In 1892, German biologist Hans Driesch isolated blastomeres from sea urchin embryos at the two- and four-cell stages, and split the cells apart. He was able to grow each cell into a fully developed individual, though it remained smaller than normal. This directly contradicted Roux's findings and demonstrated that the early blastomere cells each contain all the information needed to build the entire organism.

By the start of the 20th century, all was in place for a full understanding of generation and embryology. The roles of the egg and sperm had been established, and the development of the embryo by dividing cells had been observed. Spontaneous generation had been dismissed. Discovering exactly how the blastomere could contain and operate the 'recipe' for making the organism would be the work of the next century, after the revolution in genetics.

CHAPTER 6

The best idea
EVER

'Nothing in biology makes sense except in the light of evolution.'
Theodosius Dobzhansky (1973),
US evolutionary biologist

There are two possible conclusions we can draw on surveying the natural world: either it has always been like this, or it has become like this. If it has always existed in its current form, then the differences between organisms are fixed and stable. But if it came into being gradually, the whole scenario is far more interesting. Where did the organisms come from? How and why did they change? What was the natural world like before? How are organisms related to one another? When and how did they start? Are they still changing? Where will it end, if at all?

Evidence of a biological relationship between humans and other primates offended people when pointed out by early evolutionists.

In the beginning . . .

Many societies have had creation myths that explain how the world, and in particular humanity, came into being. What distinguishes these from scientific theories about the origins of the world, life and humanity is that they must be accepted as true with no proof. Indeed, proof and faith are polar opposites, since faith is demonstrated in believing things for which there is no proof. Those who accept a creation myth trust it through personal conviction rather than empirical evidence.

Science, on the other hand, is not about belief. Scientists look for evidence, build theories and test them. One theory is replaced by another, or refined, as new evidence comes to light. If a better explanation is found, scientists discard the original theory and embrace the new – but on the same footing, that it can be replaced.

The ubiquity of religious belief in the West kept the origins of life from scientific scrutiny until the 19th century. Even then, the challenge to the Biblical account of Creation was contentious and succeeded only when evidence mounted to the point at which it could no longer be ignored. At last, people found a way of accommodating the scientific explanation within their religious framework; some discarded the framework.

Stability versus change

A crucial difference between Creation stories and the theory of evolution is that generally in Creation stories, organisms exist ab origine – they all appear together and remain stable. In evolutionary theory,

FROM 'UR-SLIME' TO MULTICELLULAR ORGANISMS

The German naturalist Lorenz Oken (1779–1851) proposed that life first emerged as 'infusoria' emanating from a primitive mucus-like slime which he called 'urschleim'. He claimed that all the more complex life-forms emerged from aggregates or colonies of these single-cell infusoria. Oken had no evidence for his theory – only a good imagination. But it is not very far from current thinking. The earliest multicellular life-forms were probably formed as single-celled cyanobacteria came together, acted symbiotically and eventually resolved into a group of differentiated cells.

Right: Infusoria are microscopic aquatic organisms.

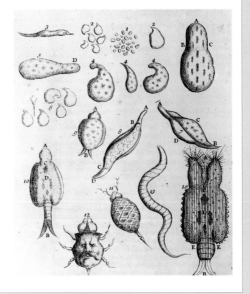

organisms evolve from physiologically simpler ancestors. The challenge to Creation mythologies has happened more than once.

From supernatural to natural

The first person we know to have rejected a supernatural explanation for the origins of life was the Ionian Greek philosopher Thales (*c*.624–546BC), who considered water to be the origin of all things. A contemporary of his, Anaximander (*c*.611–547BC), proposed a sort of proto-evolutionary theory in which animals were formed from the bubbling mud of the early Earth and lived first in the water. As the land and water separated, some of these creatures adapted to – and colonized – land. Even humans, in Anaximander's view, had developed from fish-like earlier animals.

Xenophanes (576–490BC) noticed the presence of fossils of sea creatures far inland, where they were often found in mines. He recognized them as relics of once-living animals and deduced that the land and sea had changed places in the past and would doubtless do so again in the future. He saw fossils as evidence supporting the theory

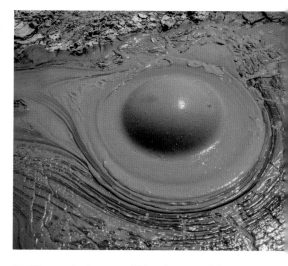

Bubbling mud volcanoes could have been one of the sites of early life.

that the Earth had once been covered with mud. Fish and other animals had been pressed against the mud and left their impression, which hardened as the mud dried out. Unfortunately, this astonishingly accurate insight would be lost for more than 2,000 years.

Empedocles (*c*.490–430BC) suggested a bizarre scenario that has a whiff of natural selection about it. He envisaged a world

in which disembodied bits at first wandered about: 'faces without necks, arms wandered without shoulders, unattached, and eyes strayed alone.' Next, all kinds of potential creatures were made up from assorted bits and pieces and only those that had a functional body plan survived: 'Clumsy creatures with countless hands. Many creatures with faces and breasts looking in different directions were born; some, offspring of oxen with faces of people . . . and creatures in whom the nature of women and men was mingled, furnished with sterile parts.'

By chance, some combinations turned out to be viable organisms. These survived, but others perished 'and are perishing still'. It's easy to see this as a very early formulation of random mutation and the doctrine of 'survival of the fittest'.

Stifling evolutionary thought

Although it might look as though the Greeks were moving towards a model that could suggest evolution, it was not to be. Two gargantuan intellects of Classical Greece, Plato and his pupil Aristotle, derailed the evolutionary train of thought.

For Plato, everything we observe in the real world is simply an imperfect echo of an

According to Xenophanes, this fossilized fish would have been created when a dead fish was pressed into mud that then hardened. He wasn't far wrong.

essential 'form' – the perfect idea of a tree, a whale, a charitable act, a song and so on. These forms are not directly accessible to us, but are immutable and perfect. The variation we see in the world around us is simply a collection of imperfect copies of the forms. This leaves no scope for evolutionary change (everything is fixed) or productive variation (since variety is failure).

Aristotle saw everything as purpose-driven, with biological development directed towards some ends. He considered that if an animal had long legs, it was because it needed to be tall. He also considered the nature of all organisms to be fixed for all time – species did not appear, develop or disappear.

Which came first, the watch or the watchmaker?

Aristotle's view was teleological: organisms have the features they need in order to perform their functions. This poses the question of who or what set the purpose. Similarly, the 'intelligent design' argument proposes that the variety and complexity

Do ant-eaters exist to eat ants or do ants exist to feed ant-eaters? A teleological view of the natural world has some tricky questions to answer.

of the natural world can only have come about through design, not by accident. The English clergyman and philosopher William Paley (1741–1805) put it succinctly when he said that if we see a complex mechanism such as a watch, we deduce the existence of a watchmaker, and it is the same with Creation. The Roman author Cicero made the same comparison with a sundial in AD45. The intelligent designer need not be supernatural; it could be an extra-terrestrial or even extra-universal intelligence setting up a science experiment on Earth.

Intelligent design is a specious argument. It accepts the world as it is and wonders at how well everything works together and is adapted to its purpose, whereas in fact the system is simply the result of the process. Organisms could all have evolved completely differently and look just as well 'designed' since the system would still work, albeit differently. The very fact that we are present to observe the outcome of the process means it has worked and so will inevitably look well-designed. On a planet in which the process did not work, there would be no observers to comment on the poor quality of the design.

The dawn of Creation

As we have seen, the dominant model of order in the natural world inherited from Classical antiquity was that of the ladder or chain of being, with every organism occupying its correct niche in a comprehensive hierarchy of life. This left no room for change or evolution, though there were occasional solitary voices of dissent (see box, left).

In the Western world, the coming of Christianity stifled inquiry into the origins of the natural world. The Bible said

> 'Why do the waters give birth also to birds? Because there is a family link between the creatures that fly and those that swim. In the same way that fish cut the waters, using their fins . . . so we see birds float in the air by the help of their wings . . . their common derivation from the waters has made them one family.'
>
> Basil of Caesarea (AD329–79)

that God created all organisms over a period of a few days; they didn't change. The theologian and philosopher Thomas Aquinas (1225–74) put in a plea for reason, pointing out that it is not essential to take every word of the Bible – or at least of the Old Testament – literally: 'The manner and the order according to which creation took place concerns the faith only incidentally.' Still, the notion that the world is unchanging, or at the very least slowly winding down from a lost golden age, left no room for evolutionary thought. Any change was in a backwards direction; it didn't involve organisms becoming better adapted to their environments.

A changing view of change

At last, Europe began to challenge Classical authority and even to query some of the implications of a literal reading of the Bible. One early proponent of a different approach to the natural world was French philosopher Pierre de Maupertuis (1698–1759).

Men from mud?

As Maupertuis was not a practical scientist he worked entirely theoretically, developing his ideas from contemplating readily available information. Even so, he seems to have anticipated the later theories of variation by mutation. In his book *The Earthly Venus* (1745), he first apologized for the potential offence that his readers might perceive: 'Do not be angry if I say you were a worm, or an egg, or even a kind of mud,' he wrote.

Maupertuis argued against both ovist and spermist views of generation, suggesting that as offspring can resemble either parent and frequently mix features of both, it is unreasonable to suppose that only one parent provides the blueprint for the offspring.

St Thomas Aquinas was a great proponent of Aristotle's works and ideas, and instrumental in reintroducing Aristotle to Western Europe.

Maupertuis believed that new varieties could arrive by chance, but he didn't rule out the impact of external conditions such as climate or food supply. He suggested an experiment, later carried out by August Weismann (*see* pages 137–8), to mutilate an organism over generations to discover whether eventually the progeny would produce the mutilation from birth. More important than any of his ideas is the fact that Maupertuis considered that the natural world can and does change.

Pierre de Maupertuis was an early champion of change in organisms over extended periods of time.

The general tenor of the age, though, remained with the idea of a fixed Creation well into the 18th century. Linnaeus (*see* pages 31–3), writing in the 1740s, was aware that variation between individuals occurred, and that some types of plant seemed to 'degenerate' into others, but he was still firmly of the view that species were essentially fixed and 'no new species are produced nowadays'.

Similar, but different

But the ball that Maupertuis had set rolling slowly gathered speed. The 18th-century French naturalist Georges-Louis Leclerc was influenced by Maupertuis' thinking. He published a colossal survey of all natural history, *Histoire Naturelle* (1750), in which he noted that the large tropical fauna of the Americas and of Asia and Africa were substantially different even when they lived in similar climates. Climate, then, could not be the only determinant of form. While the large animals of the north (elk, moose, reindeer) were similar in North America and Europe, those of the tropical countries were very different. The giraffe, zebra and lions of Africa, he noted, bear only a passing resemblance to the llamas and jaguars of South America.

To account for this, Leclerc suggested that animals had been created at the North Pole during a time when this region was warm, and had spread out, adapting as they moved southwards. Leclerc identified 38 quadrupeds from which he claimed all the others developed, and even suggested that humans and apes might have a common ancestor. He aroused the suspicion of the Church, which felt some of his ideas were bordering on heretical, and he had to be circumspect – even ambiguous – in the way he presented them.

Leclerc changed his views over the course of his life, too. At one point he ridiculed the idea of species as laid out by Linnaeus, saying there are only individuals. Only by making ever finer distinctions between 'species' can we approach truth, as this gets us closer to looking solely at individuals. He changed his mind after contemplating the problem of the sterile mule, eventually accepting that the ability of members of a species to interbreed was the true defining characteristic of species.

Leclerc thought initially that species were fixed for all time, but his discovery that imperfect, redundant or rudimentary organs in animals led him to ask why a perfect Creator would have made such things. He fudged the issue by stating that each species was originally a perfect creation, but had degenerated over time.

Towards a theory of evolution

Even change in the form of gradual degeneration marked a significant move away from the mainstream position that God's Creation was fixed for all time. Not long after, the idea of change in the other direction – improvement – was mooted.

Erasmus Darwin, grandfather of the naturalist Charles Darwin (*see* page 147), was a physician by trade, but also a polymath with formidable knowledge of natural history. He proposed a form of evolutionary theory in *Zoönomia, or The Laws of Organic Life* (1794–6). In this, he described life evolving from a single common ancestor, forming 'one living filament' – an idea with clear correlations in his grandson's work 60 years later. Erasmus Darwin tackled the question of how species could develop from one another, and considered how competition and sexual selection might direct their development: 'The final course of this contest among males seems to be, that the strongest and most active animal should propagate the species which should thus be improved.'

Despite Erasmus Darwin's forays into the territory, it was the French naturalist Jean-Baptiste Lamarck who proposed the first true evolutionary theory and, of the two, his influence was the more substantial. He proposed that species evolved over

'Would it be too bold to imagine, that in the great length of time since the earth began to exist, perhaps millions of ages before the commencement of the history of mankind . . . that all warm-blooded animals have arisen from one living filament, which the Great First Cause endued with animality, with the power of acquiring new parts, attended with new propensities, directed by irritations, sensations, volitions, and associations; and thus possessing the faculty of continuing to improve by its own inherent activity, and of delivering down those improvements by generation to its posterity, world without end!'

Erasmus Darwin, *Zoönomia* (1794–6)

time, in a smooth continuum and following natural laws. All species, even humans, had evolved from earlier species, he said. The driving forces of evolution were, for him, the rather romantically named '*le pouvoir de la vie*' ('life force', or 'power of life') and '*l'influence de circonstances*' ('the influence of circumstances'). The former drove organisms to become ever more complex (human beings having clearly achieved this most dramatically), while the second accounted for adaptability to the prevailing environment. This gives evolution a similar hierarchical structure to that of the Great Chain of Being (*see* pages 24–6), with the most complex organisms considered the pinnacle of evolution and superior to those that have simpler systems and bodies, even if the simpler organisms are more successfully adapted to their environment.

CATASTROPHISM

While Lamarck proposed a process of smooth evolution over an extended timescale, the French anatomist Georges Cuvier (1769–1832) favoured catastrophism, proposing that change came about as a result of intermittent catastrophes – now called 'extinction events'. He came to this conclusion from his work on geology, noticing that the fossils of prehistoric animals such as mastodon (a relative of elephants) and megatherium (a giant sloth) are very different from their modern counterparts.

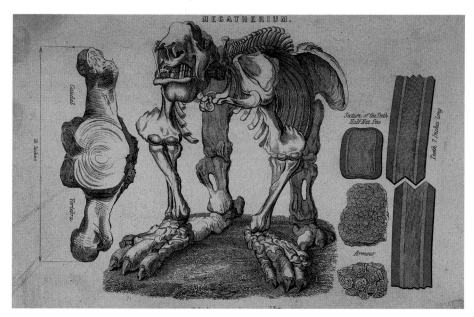

JEAN BAPTISTE PIERRE ANTOINE DE MONET, CHEVALIER DE LAMARCK (1744–1829)

Despite his impressively long noble name, Lamarck struggled with poverty throughout his life. He was the youngest of 11 children, but as all his brothers died before their father he inherited the family title – and just enough money to buy a horse. Having been forced by his father to train for holy orders, he abandoned his studies immediately on his father's death, bought a horse and left to join the German army. A probably exaggerated account records that all the officers in his company were killed in battle and he assumed command. He was so brave that he was immediately promoted, but at 22 was forced to leave the army because of ill-health.

He spent the rest of his years living in Paris, barely scraping a living as a bank clerk while also studying botany, medicine and music. Helped by Leclerc, who was impressed by his work on French plants, *Flore français* (1778), Lamarck was appointed assistant to the botanical garden of the royal natural history museum and travelled around Europe collecting specimens.

With the removal of so many scientists in the French Revolution, Lamarck was appointed to a chair in zoology (about which he knew little) and re-educated himself as a specialist in invertebrates. Despite his progress, he remained poor, and suffered considerable personal losses: he was widowed four times, most of his children predeceased him, in old age he went blind and on his death he had only a pauper's grave, his bones tossed into a trench with others.

Long necks and no tails

Lamarck is often mocked for the idea that changes or adaptations acquired during an organism's lifetime can be passed on to subsequent generations and thus direct the path of evolution. A typical example of a Lamarckian explanation is that giraffes strain upwards, stretching their necks to reach high leaves. Over time, the giraffe's neck elongates and later giraffes are born with longer and longer necks. The standard (accepted) evolutionary explanation would be that in the variety of giraffes born, those with longer necks are more successful at reaching vegetation which is out of reach to other animals (including shorter giraffes) so they are more likely to survive, breed and pass on their long-necked genes. The appearance of the long neck is one among many variations in a population of organisms, but as it aids the animal's success, natural selection favours it. It's not the same as giraffes stretching their necks and giving birth to long-necked progeny as a result.

There were, indeed, scientists who experimented by altering animals and testing the heritability of the change, but these bore little resemblance to Lamarck's thinking. Lamarck never suggested that deliberate mutilation, or the results of accident or illness, would lead to adaptation in future generations.

The German biologist August Weismann (1834–1914) cut the tails off generations of mice, finding the progeny again and again born with intact tails. Of course, his experiments never yielded naturally tail-less mice – which was what he hoped to show. Weisman's experiment supported his theory of 'germ plasm' and the Weisman barrier – the notion that heredity occurs through the germ cells (eggs and sperm) alone and that changes to body cells will not be passed on to the next generation.

A giraffe's long neck could be the result of intelligent design (the Church) or generations of striving by its ancestors to reach foliage at height (Lamarck) or the success of its ancestors who happened to have longer necks (Darwin).

LAMARCK HAS THE LAST LAUGH?

Lamarck's idea that the behaviour and experiences of one generation might affect the development of future generations of organisms has found some favour again in the late 20th and early 21st centuries, with the science of epigenetics (the control of gene expression) and the discovery of some genuine epigenetic effects. It appears that when people (and presumably other organisms) are subjected to environmental stress it can modify how their genes are expressed and this can be passed on to later generations. Some environmental stresses have the effect of turning on or off the expression of certain genes in at least the next one or two generations.

One piece of evidence for this is found in the descendants of victims of the Dutch 'Hunger Winter', a famine that occurred in 1944, during World War II. Children conceived towards the end of the famine, who had been undernourished in the womb for the first three months of gestation, were born at a normal birth weight but had enduring problems thereafter, including a greater than normal tendency to obesity and to cardiovascular disease. The problems persisted into the next generation: the children of these gestationally malnourished infants also had a greater propensity to obesity and some other health problems. The epigenetic explanation is that the stress of the famine *in utero* altered the way genes were expressed that remained for the whole of the individual's life and, apparently, affected gene expression in their offspring.

Although Lamarck's explanation of evolutionary change is no longer accepted, he was innovative in proposing a mechanism. His idea that a changing environment prompted adaptation in organisms is also part of later evolutionary theory, though no longer effected through the organism's need and striving.

> 'And striving to be man, the worm
> Mounts through all the spires of form.'
> Ralph Waldo Emerson, 1836

The evidence is under your feet

The 19th century was a time of extraordinary advances in geology which also had a huge impact on the story of biology. The best evidence we have of extinct organisms – for there having once been completely different life-forms – is the fossil record.

Stones or bodies?

People have been finding fossils for millennia, and have explained them in various ways. In China, dinosaur fossils were sometimes thought to be dragon bones, for example. There were two explanations for fossils: they are the remains of once-living creatures; or they formed as fossils, and were never alive.

An Ancient Greek theory, dating back at least to the time of Aristotle, was that fossils formed within the Earth and grew in rocks as the result of some shaping force. This 'extraordinary Plastick virtue' made stones that had never lived look like living things.

The alternative view was that some process changed once-living organisms into stone after their death. This was accounted

A Tyrannosaurus rex *fossil prior to excavation – almost incontrovertible evidence that creatures which are now extinct preceded us on Earth.*

for by the action of 'petrifying fluid', as described by the Arab scientist Ibn Senna (c.980–1037), better known today as Avicenna. The philosopher Albert of Saxony (1320–90) popularized the idea in Europe.

The locations at which fossils were found could also pose a puzzle. The Ancient Greek philosopher Xenophanes (570–480BC) concluded that some areas that were dry land in his day must previously have been under water because fossilized sea shells were found there. The 11th-century Chinese naturalist Shen Kuo noticed that fossilized bamboo was found in areas that were then too dry for bamboo to grow. As an explanation, he proposed that the climate had changed since the distant past. The discovery in the Middle Ages and later of fossilized sea creatures in rock beneath Paris challenged belief in an unchanging world. One explanation was that they were the remnants of meals thrown away by travellers and had petrified relatively recently.

From dragons to dinosaurs

As long as fossils were discovered which looked sufficiently like still-living organisms, it was possible to believe in a 'petrifying' solution without having to worry about changing forms, evolution or extinction. But when fossils were found that could not be matched to current life-forms, it raised problems.

Only one explanation would fit the evidence – that some types of organism which had once lived were no longer found on Earth. Empedocles (490–430BC) suggested this possibility – that animals might not be immutable, but that they could change, some surviving and others perishing over time.

It was not a popular view, though, particularly in light of the tradition within the Abrahamic religions that God had created all animals and plants in discrete and finished forms. If God had created a perfect world, how could some animals die out over time? It didn't fit the paradigm, so for a long time was not considered a possible explanation. It was not until substantial numbers of fossils were unearthed in the 19th century that the question became a serious issue.

With advances in microscopy, the evidence that fossils had once been living creatures grew to the point where it could no longer be ignored or explained away. In *Micrographia*, Robert Hooke (*see* pages 88–90) compared the microscopic structure of fossil wood with ordinary wood. He concluded that petrified wood and fossil shells, such as the plentiful ammonites, were the remains of living organisms which had been transformed by soaking in 'petrifying water' that contained dissolved minerals.

As fossil organisms found in rock were often very different from anything living in

> 'Our English Quarry-shells were not cast in any Animal mold, whose species or race is yet to be found in being at this day . . .
> I am apt to think, there is no such matter, as Petrifying of Shells in the business . . .
> but that these Cockle-like shells ever were, as they are at present, lapides sui generis ['stones of their own kind'], and never any part of an Animal.'
> English naturalist Martin Lister, 1678

the 17th century, Hooke came to the unpopular conclusion that some species alive in the past had since become extinct, perhaps as the result of a geological catastrophe. This still did not suggest that any organism had changed in form, but it was one heretical step closer to evolution.

The English naturalist John Ray rejected Hooke's idea of extinctions, considering it theologically unacceptable. However, when the English lawyer and geologist Charles Lyell (1797–1875) read Hooke's account nearly 200 years later, he called it 'the most philosophical production of that age, in regard to the causes of former changes in the organic and inorganic kingdoms of nature' (1832).

The Earth moves

The ink was barely dry on *Micrographia* when, in 1666, two Italian fishermen caught a giant shark, which was sent to the Danish anatomist Nicolas Steno (1677–86) for dissection. Steno noticed the similarity between the shark's teeth and the so-called tongue stones often found in rocks. He concluded that tongue stones were, in fact, shark teeth that had turned to stone over a long period of time. To explain it, he proposed that rock was originally molten and was laid down in layers, sometimes trapping an animal that then turned to stone as its bodily 'corpuscles' were exchanged for corpuscles of rock. He declared that the oldest layers of rock were lowest and the newer layers uppermost, establishing for the first time the chronological principle of geology known as Steno's 'Law of Superposition'.

NOAH'S ARK

For those who believed in the literal truth of the Bible, the account of Noah's ark was further evidence of the stability and enduring nature of species.

'Buteo and Kircher have proved geometrically that, taking the cubit of a foot and a half, the ark was abundantly sufficient for all the animals supposed to be lodged in it . . . The number of species of animals will be found much less than is generally imagined, not amounting to a hundred species of quadrupeds.'

Encyclopaedia Britannica (1771)

'However trivial a thing a rotten shell may appear to some, yet these monuments of nature are more certain tokens of antiquity than coins or medals, since the best of those may be counterfeited or made by art and design, as may also books, manuscripts, and inscriptions, as all the learned are now sufficiently satisfied has often been actually practised . . . though it must be granted that it is very difficult to read them and to raise a chronology out of them, and to state the intervals of the time wherein such or such catastrophes and mutations have happened, yet it is not impossible.'

Robert Hooke, *A Discourse on Earthquakes*, 1705, published posthumously

The fact that Hadrian's Wall between England and Scotland has not fallen down as a result of geological movement demonstrates that such movement is very slow.

A hundred years later, the Scottish geologist James Hutton (1726–97) studied the rock formations of his native land and published his findings in *Theory of Earth* (1788). The book was groundbreaking. Hutton considered the Earth to be millions of years old – far older than the 6,000 years calculated by Bishop Berkeley from Biblical genealogies. Hutton described gradualism, by which the processes of uplift and erosion are slowly and continually shaping the world as they have done for a very long time. Citing as evidence the lack of change to Hadrian's Wall, a structure built by the Romans 1,500 years previously, he demonstrated that geological change must be very gradual, so the age of the history of the Earth must extend far back into the distant past.

Charles Lyell's *Principles of Geology* (1830–33) further developed gradualism, then known as uniformitarianism, which stressed the consistency of the processes of change. This might seem contradictory, but it is a crafty answer to the theological conundrum of a world that was providing increasing evidence of long-term change while the Church still wanted to believe that God had fixed Creation for all time. The work-around was that the laws of nature were fixed, and they dictated the pattern of steady, ongoing change. Darwin read a copy of Lyell's work on his famous voyage of discovery aboard HMS *Beagle*, and sent back many reports of geological features he had seen which supported Lyell's theories.

Fossils too big to ignore

It was one thing to discover small shells and debate whether one type of shellfish had degenerated into another, but once large fossils began to emerge it became increasingly difficult to sustain the view that God had created a basically stable and unchanging world.

In 1676, a large bone was recovered from a limestone quarry in Oxfordshire and sent to Reverend Robert Plot (1640–96), an English naturalist and the first curator

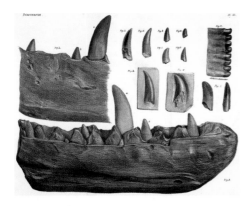

of the Ashmolean Museum in Oxford. The fragment was probably a *Megalosaurus* femur. Plot's religious position did not allow him to think of huge, no longer extant animals, so he concluded that it must have come from a giant human killed during Noah's flood. (The Bible allows for the existence of early giants.)

But the evidence built up. More and more fossils were discovered in the late 18th century. In 1815, bones said to be from a giant tetrapod were passed to William Buckland (1784–1856), Professor of Geology at Oxford University. Cuvier, the French anatomist, visited Buckland in 1818 and realized the bones were from a very large lizard-like creature. In 1821, Buckland and the English palaeontologist William Conybeare studied the bones in detail, referring to them as the 'Giant Lizard'. The knife-like teeth had clearly once belonged to a large predator. There were no large predators like this in England, so Buckland had to admit they came from something that no longer existed. He named it *Megalosaurus* in 1824, and gave the first account of a dinosaur (though he did not use the word).

Dinosaurs take off

Dinosaurs came to light thick and fast in the 19th century. The limestone quarries of Oxfordshire continued to be productive, and the cliffs around Lyme Regis produced more fossils of plesiosaurs and ichthyosaurs. William Buckland also found remains of

Above left: The jaw and teeth of a Megalosaurus *clearly indicate a large, carnivorous lizard-like creature.*
Bottom left: Iguanodon *and* Megalosaurus *as imagined from fossil evidence in the 19th century.*

THE SCROTUM DINOSAUR

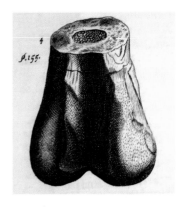

Robert Plot published a description of his piece of femur along with an illustration. The illustration was reused in 1763 by the English naturalist Richard Brookes (1721–63). He thought it resembled a pair of human testicles and so called it 'Scrotum humanum'. If Brookes actually intended this as a name rather than a descriptive label, the name *Megalosaurus* should really have been dropped in favour of *Scrotum humanum*, as this took precedence. Biologists have found many reasons not to rename the dinosaur, the most convincing of which is that since the fragment has been lost it can't be conclusively proven to be from *Megalosaurus*. But it's tempting to think they just don't want the first dinosaur named after the scrotum.

The end of the Megalosaurus *femur looks uncannily like a human scrotum.*

MARY ANNING (1799–1847)

English palaeontologist Mary Anning began fossil-hunting as a child with her father and brother. They had a profitable stall selling the fossils, which were easy to find in the fallen cliffs of Lyme Regis. The fossils had local names, such as 'snake-stones' (ammonites), 'devil's fingers' (belemnites) and 'verteberries' (vertebrae). In 1811, when Mary Anning was 12 years old, her father found what turned out to be an ichthyosaur skull. Mary found the remainder of the skeleton a few months later, the first ichthyosaur ever discovered. She went on to find the first plesiosaur fossils, and the first pterosaur fossil discovered outside Germany.

Mary became the most proficient of the family fossil-hunters, finally running the family business and opening a shop, Anning's Fossil Depot. This attracted rich and powerful customers from around Europe and even America, including King Frederick Augustus II of Saxony and buyers for important museums. Anning educated herself thoroughly about fossils and the anatomy of extant animals, dissecting fish and cuttlefish so that she could better understand the structure of the fossils she found. But Anning was treated badly by the scientific community. She knew more than the educated male geologists who consulted and bought from her, and they frequently published information she had given them without acknowledging her.

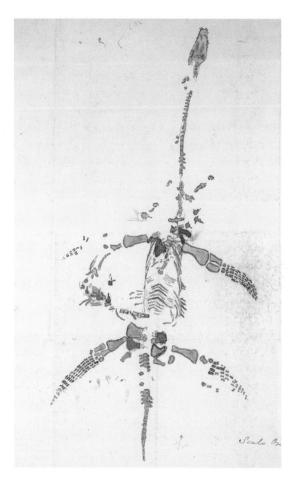

A plesiosaurus drawn by Mary Anning, from her letter describing its discovery.

was a difficult character, often labelled cantankerous, and had frequent spats with other palaeontologists and geologists. A committed Creationist, he took issue with early evolutionary ideas ('transmutationism') and later with what he considered Darwin's assault on the Bible narrative. His underhand methods (such as writing anonymous attacks on rivals' work) made him unpopular. He held the view that dinosaurs were not large reptiles, but had mammal-like features given to them by God, and were more like elephants or rhinos. They could not have transformed from reptiles, he felt sure, so there need not be any massive challenge to the established Creation narrative.

After *Megalosaurus*, the next large dinosaur to come to light was *Iguanadon*, discovered by the English physician and palaeontologist Gideon Mantell (1790–1852). He had found teeth from a large herbivore and parts of 'an animal of the Lizard Tribe of enormous magnitude' in the early 1820s. The teeth were dismissed by Buckland as belonging to either a fish or a rhinoceros, and by Cuvier in Paris as rhino teeth. But then Cuvier immediately changed his mind, deciding they were indeed reptilian; he proposed some giant reptilian herbivore.

Mantell tried to find a modern reptile with comparable teeth. Samuel Stutchbury, an assistant curator at the Royal College of Surgeons, pointed out that they resembled

mammoths, coprolites (fossilized faeces) and remains of an early human, now dated to 31,000BC. It soon became apparent that *Megalosaurus* was far from the only early giant.

The name dinosaur, meaning 'terrible lizard', was coined by the English naturalist Richard Owen (1804–92) in 1842. Owen went on to found the British Natural History Museum, opened in 1881. He

> *'I used to be ashamed of hating him so much, but now I will carefully cherish my hatred and contempt to the last days of my life.'*
>
> Charles Darwin on Richard Owen

iguana teeth, but were 20 times larger – so the animal found its name, *Iguanodon* or 'iguana-tooth'. Although Owen tried to persuade Mantell that *Iguanodon* must have been a heavy, lumbering mammal-like creature, Mantell realized a little before his death that it had slender front limbs. Mantell died, though, and Owen's interpretation won the day and was accepted for a considerable period.

Where did they go?

Perhaps Cuvier's most important single contribution was the theory known at the time as 'catastrophism', which is now considered the first suggestion of mass extinctions. In 1813 he proposed that instead of the very slow and steady changes described by gradualism, the Earth was occasionally subject to catastrophic change in the form of major floods. After these, new organisms appeared – but by a process of creation, not evolution. He was highly critical of Lamarck's ideas about animals changing over time.

Cuvier is often considered the father of paleontology, and helped to establish comparative

Gideon Mantell's reconstruction of Iguanodon *put the thumb spike on the nose as a horn.*

anatomy. He was the first to try to include extinct, fossilized animals into Linnaean taxonomy. By establishing extinction as a fact, he facilitated a great step towards evolutionary theory.

Evolution – now with dinosaurs

It's impossible to imagine the impact that recognizing dinosaur fossils must have had in the 19th century. It was a paradigm shift. It now appeared that the Earth was considerably older than 6,000 years and that unimaginable monsters, now clearly extinct, had once walked its surface. Fossil-hunters noted that no human remains were found in the same strata as the dinosaurs. The truth was becoming impossible to resist: the Bible hadn't got it right – not even close. For Creationists such as Owen, fudging the issue was an attractive option, but it wasn't going to be sustainable. It finally became untenable in 1859.

A man, a Beagle and some finches

The story of evolution, of course, comes to a climax with the work of Charles Darwin. But it was a climax that was a long time coming. Darwin was only 22 when he

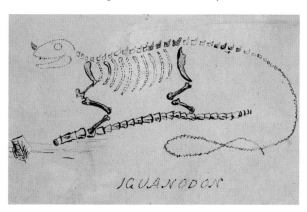

IGUANODON

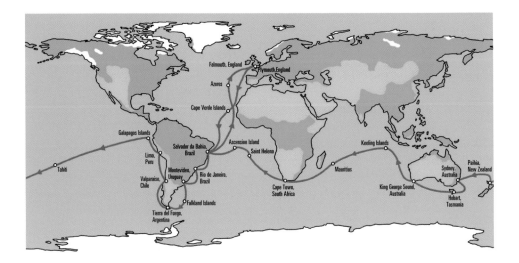

was offered the position of naturalist on the survey ship HMS *Beagle*. He was not first choice, and it was only because others had professional or domestic responsibilities that he was asked at all. He was almost refused by the captain, Robert FitzRoy, in part because FitzRoy thought the shape of his nose indicated weakness. But with all problems solved (except the shape of his nose), Darwin set sail in late December 1831 on a journey that would take him around the world and keep him from home for nearly five years.

Darwin saw volcanoes, earthquakes, rainforests, mountains, the wide oceans, the most beautiful landscapes on Earth, and people and animals of every variety. Most importantly, he collected thousands of biological and geological specimens.

Shells collected by Darwin during his voyage on HMS Beagle.

The route of the Beagle (*in red*) *took Darwin to South America and the Galapagos Islands, but also to Australasia and South Africa.*

Darwin read Lyell's geology text and looked everywhere for evidence of the geological history of the world and endorsement of Lyell's theories – and he found it. He considered the placement of sea fossils on dry land and on mountains; he pondered the fossils of large extinct mammals; he noted the rising of coral atolls; and, most importantly, he noticed and puzzled over

'THE GREATEST ENGLISHMAN SINCE NEWTON' – CHARLES DARWIN (1809–82)

Like many people who later achieved great things, Darwin was unpromising at school. He was prone to performing pranks and avoiding work, activity that caused his headmaster at Shrewsbury school to call him a wastrel and his father to predict that 'you will be a disgrace to yourself and all your family'.

After school, Darwin began to study medicine at Edinburgh University, but found it both boring and disgusting. He moved to Cambridge to train for the clergy. That was no more successful: he skived lectures, and spent his time hunting and drinking. But he'd always had an interest in the natural world, and this was nurtured at Cambridge through his friendship with a highly respected professor there, the English clergyman and botanist John Stevens Henslow (1796–1861). Darwin spent so much time with the professor that he was known to the dons there as 'the man who walks with Henslow'.

At Edinburgh he had learned from a freed Guyanan slave, John Edmonstone, how to stuff birds. At Cambridge he developed a fascination with beetles and had also begun to study with the celebrated English geologist Adam Sedgwick (1785–1873), spending two weeks with him studying rock formations in Wales. He was planning a trip to study the flora and fauna of Tenerife when he received a letter from Henslow proposing him as naturalist on HMS *Beagle*, sealing his future and the history of biology. After nearly five years, Darwin returned to England with an established reputation as a naturalist. He was invited to join all the leading scientific institutions of the day and worked alongside the most learned minds while at the same time developing – at first haphazardly – his theory on evolution.

In 1839, after debating with himself whether marriage would be of benefit, he finally married his cousin, Emma Wedgwood, having come to the conclusion that a wife would be 'better than a dog anyhow'. In fact, Emma proved a devoted companion and stayed with him until his death 48 years later – which was indeed longer than a dog would have lasted.

Darwin largely stayed out of the uproar that followed the publication of his groundbreaking book, *On the Origin of Species* (1859). He remained intellectually active, publishing until his death. In a strange coincidence, his final work on earthworms was co-authored with the grandfather of Francis Crick, one of the discoverers of the structure of DNA (*see* pages 164–9).

the distribution of different types of animals, particularly the strange, isolated species found on some islands. He made copious notes, but none hint at the theory which would develop later from his observations.

On his return, Darwin was fêted. He was friends with the most brilliant minds of his generation. His reputation was secured by his collected samples and his long account of the voyage of the *Beagle*, published in 1839. In private, he continued to write notebooks and think, slowly working out the implications of all he had noticed over his long period away.

Evolution evolves

On the issue of evolution, Darwin originally agreed with Lyell, who proposed that species had diversified from various 'centres of Creation'. And this had remained his view on the *Beagle*. But then Darwin began to think in terms of a 'tree of life', a phrase that appears with increasing frequency in his notes, and he decided that Lamarck was

THOSE FINCHES

Darwin is most often associated with the Galapagos Islands and the finches with differently shaped beaks that he found there. He was in the Galapagos for only a few weeks, and his notes from the time do not make much of the finches. At the time, he was more interested in the giant tortoises and the way reptiles had occupied the niches that in most ecosystems go to mammals. Darwin donated the birds he collected to the Royal Geological Society, and they were passed on to the English ornithologist John Gould (1804–81) to study. It was Gould who recognized that the 'blackbirds', 'gross-beaks', 'warblers' and 'finches' that Darwin had collected were all descended from the same species of South American ground finch. The way the beaks of the finches had become adapted to the foods available on the different islands proved to be a perfect demonstration of adaptation and speciation. The birds probably came

from mainland Ecuador, but once separated between the islands they had developed into independent colonies, each evolving separately to occupy the various ecological niches they found there. Some evolved small beaks suitable for eating seeds, for example, while others evolved sturdy beaks which could crack nuts.

The beaks of the Galapagos finches, each suited to the diet available to it.

> 'Mention persecution of early Astronomers.— then add chief good of individual scientific men is to push their science a few years in advance only of their age. (differently from literary men.—) must remember that if they believe & do not openly avow their belief, they do as much to retard, as those, whose opinion they believe have endeavoured to advance cause of truth.'
>
> Charles Darwin, Notebook C, 1838

wrong to think that change came about as a result of an organism's striving, but that it came instead from adaptation. He did not entirely reject the divine, writing: 'Let animals be created, then by the fixed law of generation, such will be their successors.'

By the late 1830s, he was referring frequently to 'my theory' – it seems it was taking root, but he was not yet ready to go public with something that would undoubtedly cause an almighty rumpus. He began work on formulating his theory in 1837, but did not finish *On the Origin of Species* until 1858. It was published the following year. It has been traditional to suppose that Darwin delayed publication because of fear. Though he doubtless anticipated problems, there is no evidence he delayed publication rather than just worked for a long time on his book – an entirely appropriate thing to do given the revolutionary nature of his thesis and the fact that he also had other work to contend with. Indeed, he spent eight years studying

> 'Man in his arrogance thinks himself a great work worthy the interposition of a deity. More humble and I believe truer to consider him created from animals.'
>
> Charles Darwin, Notebook C, 1838

barnacles in the middle of writing it (he would later state, 'I hate a barnacle, as no man ever did before').

Darwin recognized the implications of his line of thinking early on, saying of animals that 'they may partake from our origin in one common ancestor; we may be all netted together.'

Darwin's theory takes shape

Darwin said he began to think clearly about evolution in 1838, the year in which he read *An Essay on the Principle of Population* by the English scholar Thomas Malthus. The essay gives a gloomy prognosis for the future of mankind, claiming that populations will always increase to the point that their food supply cannot sustain them, and then famine will inevitably cull the population. Darwin took the Malthusian idea of competition and survival and extended it to the whole of the natural world. He imagined 'a force like a hundred thousand wedges trying to force every kind of adapted structure into the gaps in the economy of nature or rather forming gaps by forcing out weaker ones.'

A nasty shock

After studying barnacles, from 1846–54, Darwin returned to formulating his ideas on evolution. He carried out experiments to test whether seeds could cross the sea and still be viable (he decided that they could), and began to put together notes for what he now realized would be a large book. In 1856, Lyell remarked that he wished Darwin would publish something before someone else came up with the idea. In 1857, that's exactly what happened. The English naturalist and explorer Alfred Russell Wallace sent Darwin a short paper on his own work on species in the Malay Archipelago. Wallace noted that this archipelago was in two distinct parts – a western region containing species of mainly Asian origin, and an eastern part with mainly Australasian fauna (*see* pages 185–6). In this paper, he suggested a mechanism that was essentially the same as Darwin's proposal of evolution by natural selection. Distraught, Darwin sought Lyell's advice. A compromise was reached in which Darwin and Wallace both published contributions to the Linnaean Society on the same day. They aroused little interest, but it was the spur Darwin needed, and he set about writing *On the Origin of Species* in earnest.

Wallace was always gracious in acknowledging Darwin's priority and gave his full support. His views were slightly different, in that he was of a spiritual bent and rejected the idea that the human mind was the result of evolution. He had a teleological view of evolution, seeing it as working towards the human, and some 'unseen world of spirit' intervening to produce the miracle which he considered the human mind.

Thomas Malthus suggested that the human population is self-limiting.

'A stumble in the right direction'?

On the Origin of Species was finally published in 1859. Its central thesis, now familiar to everyone, is that:

• Organisms are in constant competition for resources (food, living space, mates).
• Any genetic variation in an organism which is advantageous in this competition is likely to make that organism more successful.
• The successful organisms get to reproduce, so pass on their beneficial adaptations.
• A beneficial adaptation becomes increasingly common in the population until eventually it becomes a feature of that species.

Darwin called the processes at work 'descent with modification' and 'natural selection'. Descent with modification explains the slow change in organisms over time; natural selection explains why particular variations survive. There are always too many individuals born for them all to survive, so competition between individuals is fierce. The process of modification is slow, and there must be many intermediate stages.

It is, according to Darwin, unsurprising that there are few intermediates available for viewing, since the fossil record is so incomplete. He pointed out, too, the large number of extinct species of which we have some relics in the form of fossils – clear evidence of organisms which were no longer successfully adapted to their environments and so perished.

'In a drop of sea-water, we see earliest creation recapitulated. God does not work in one way today and another tomorrow. I do not doubt that my little droplet of water will, by its transformations, tell me the history of the universe. Let us wait and observe. Who can foresee the droplet's history? Animal-plant, or plant-animal: which will be the first to emerge from it? Mightn't this droplet be the infusorium, the primordial monad who, by its own vibrations, soon becomes a vibrion, who, ascending rung by rung, becomes a polyp, a coral, a pearl, and perhaps in ten-thousand years attains the stature of an insect? Will this droplet, or that which will become of it, be a vegetable fibre, a light, silky bit of down one would hardly even take for a living creature, but still, no less than the first hair of a newborn goddess, a sensitive, loving hair: the hair of Venus? This is no fable, this is natural history. This hair with two natures (vegetable and animal), the descendent of our droplet, is the ancestor of life itself.'

Philosopher Jules Michelet (1861)

The response to Darwin's book ranged from great enthusiasm to guarded respect, but it aroused curiosity and was commercially successful. In 1860, public debate grew more heated, particularly after a famous colloquium in Oxford at which the biologist Thomas Huxley defended Darwin's theory against the Creationist argument put by Samuel Wilberforce, Bishop of Oxford (aided by Darwin's nemesis, Richard Owen).

In *Origin*, Darwin steered clear of talking about the origins of life or saying much about human evolution. He went on to publish *The Descent of Man*, in which he made the link between humans and apes and proposed that humans evolved from other animals. Darwin stayed out of the fracas, but some of his supporters formed a dining club, the X Club, and defended evolution against critics. Two of the members, Thomas Huxley and the botanist Joseph Hooker, started the journal *Nature* in 1869.

Belief in evolution was quickly adopted in scientific circles, though Darwin's explanation of change through natural selection was not as widely accepted. Some scientists preferred a more Lamarckian explanation, and some supported the

The idea that humans have evolved through earlier hominids from other apes was abhorrent to many 19th-century Christians.

idea of orthogenesis – the theory that evolution progressed in a linear fashion towards perfection. This was a teleological explanation that involved some kind of (potentially divine) driving force.

Natural and artificial selection

Darwin coined the term 'natural selection' to draw a parallel with the 'artificial selection' used in plant and animal breeding. He began *Origin* by contemplating the huge variety in domesticated breeds. These

THE MODERN THEORY OF THE DESCENT OF MAN.

Evolution was still seen as progress towards 'better' or 'higher' organisms, as Haeckel's depiction suggests.

Fancy pigeon (left), and rock dove (right).

variations, he pointed out, have come about because humans have deliberately bred plants and animals to reinforce particular features. The wild rock dove and specially bred fancy pigeons are of the same species, *Columba livia*, yet look so very different that to the uninitiated they might well appear to be different species. If humans can artificially select to reinforce desirable traits, it is a small step to seeing the same thing happening through natural processes: so natural selection reinforces advantageous traits and brings about modification in organisms.

Dinosaurs – now with evolution

While Darwin was working on his book, the first of many dinosaur fossils were coming to light in North America. It began with teeth in 1855, then in 1858 the American palaeontologist Joseph Leidy described a *Hadrosaur* found in New Jersey. In 1861, the discovery in Bavaria of *Archaeopteryx*, an early transitional bird with many standard dinosaurian features, seemed to support Darwin's theory. With the expansion of the railroads and the further exploration of the west of America after the Civil War, more dinosaur fossils emerged. The Western

Interior Seaway was discovered in Kansas, a prehistoric sea that 100 million years ago split North America from top to bottom. It would eventually yield superb fossils.

US palaeontologist Othniel Marsh found further toothed birds in Kansas in 1872 and early fossil horses – both early birds and horses became important in demonstrating evolution. In fact, 1872 marked the beginning of a period known as the 'Bone Wars' in which Marsh and his adversary, fellow American Edward Drinker Cope, competed for supremacy in dinosaur hunting.

This fossil of Archaeopteryx *clearly shows wings with claws, feathers, and a beak with teeth.*

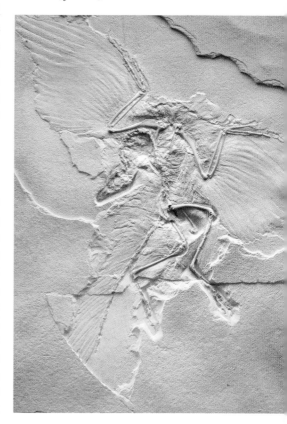

The Bone Wars

Cope and Marsh embarked on a highly competitive race to find and name dinosaurs in the last decades of the 19th century. Sometimes called the 'Great Dinosaur Rush', this was almost as unruly and unprincipled as that other great drive of the Wild West, the Gold Rush. Cope worked for the Academy of Natural Sciences in Philadelphia; Marsh represented the Peabody Museum of Natural History at Yale. Both should have known better, but they resorted to theft, bribery, poaching each other's staff, trashing each other's research and even destroying bones in an effort to undermine their rival's work and come out on top.

Between 1877 and 1892, both expended the whole of their large personal fortunes on fossil hunting in the great bone beds of Colorado, Nebraska and Wyoming. Though financially ruined, by the end they had excavated huge numbers of bones, many still unprocessed in crates when they died. They had named more than 140 species of dinosaur between them (though only a fraction of those names are still used).

One of the richest bone fields explored by the two palaeontologists was Como Bluff in Colorado. There, Marsh's men unearthed *Stegosaurus*, *Apatosaurus* and *Allosaurus*, all named by Marsh in the December 1877 issue of the *American Journal of Science*.

They made mistakes. Cope published his reconstruction of the plesiosaur *Elasmosaurus* with the head at the wrong end, and was then so embarrassed that he tried to buy up every copy of the journal in which it was published. Marsh was so slow in paying his workers that some defected to Cope. The squabbling between the two men brought American palaeontology into disrepute – perhaps even ridicule – for decades, yet their findings remain monumental. It was the Bone Wars, too, that brought dinosaurs to public attention and cemented them in our affections.

In 1878, the first of 30 fossilized Iguanodon *skeletons was discovered in a mine in Belgium. The artist Gustav Lavalette was commissioned to sketch them in their original poses before work began on preparing them.*

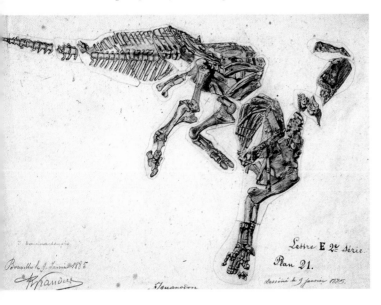

The missing link

Darwin's explanation of the gradual changes brought about in organisms as they competed with one another and adapted to their changing environment became increasingly compelling. A growing body of evidence from natural history showed that organisms do indeed change, under various types of pressure. What was missing from the picture, though, was exactly how these changes were propagated and passed on to future generations. The mechanism of inheritance remained elusive. The following century would see the pieces fall into place, as evolution and genetics came together in the so-called 'modern evolutionary synthesis'.

BARNUM BROWN AND THE T. REX

Marsh and Cope were not the only colourful figures involved in dinosaur hunting. Another American, Barnum Brown – named after a circus strongman – used to attend digs in a floor-length beaver-fur coat and a top hat. His most famous discovery was *Tyrannosaurus rex* in 1902. Like Cope and Marsh, he used dynamite to blow apart hillsides in his search for fossils, and collected so many that crates of his finds were still unexplored many decades after his death.

Barnum Brown on a dinosaur dig in Canada in 1912.

CHAPTER 7

Parents and
PROGENY

'*Heredity is brought about by the transference from one generation to another of a substance with a definite chemical, and above all molecular, constitution.*'

August Weismann, 1885

Once the idea of evolution was accepted, new puzzles arose. There must be a way for characteristics to pass from parent to progeny and also to change during repetition of that process. Unravelling exactly how evolution could happen relied on an understanding of genetics. At the very beginning of the 20th century, evolution and genetics began to come together in the most fruitful of syntheses.

Resemblances between family members are a sign of genetic inheritance that has always been visible.

The monk and the peas

While Darwin was weathering the storms of protest over the publication of *On the Origin of Species*, a Franciscan monk was growing peas in a monastery in Moravia (now part of the Czech Republic). Gregor Mendel (1822–84) carried out experiments on how pea plants inherit characteristics. Over a period of eight years, from 1856 to 1863, he grew 29,000 pea plants in the experimental gardens of his monastery, St Thomas's in Brno. He noted the patterns of inheritance of seven traits: height, flower colour, position of flowers, seed shape, seed colour, pod shape (round or wrinkled) and pod colour.

Factors and forms

Mendel discovered that certain characteristics turn up in approximately one quarter of offspring – even when they are not present in either of the parents. From the statistics, he was able to deduce the process. He began by suggesting that invisible 'factors' carry information between generations. (These factors are now called genes.) He proposed

that for each trait there are two 'forms' (now called alleles) and that they are passed on in pairs. So a factor (gene) for colour might have two forms (alleles), one of which causes a flower to be white while the other causes it to be purple. There is, he explained, a dominant form and a recessive form.

The organism will display the dominant form if it has at least one copy of that allele. It will only display the recessive form if it has two copies of that allele and none of the dominant one. If we say the dominant purple form is represented by the allele A, and the recessive white form is represented by the allele a, the allele-pairs AA, Aa and aA will all produce purple flowers, as A trumps a. Only a plant with the pair aa will produce white flowers.

Mendel's laws

Mendel formulated these findings in his 'laws of inheritance'. The Law of Segregation states that when the gametes (sex cells, sperm and eggs) are produced, the pairs of alleles split up so that each sex cell has only one for each gene. So a pea plant might pass on either an allele for white flowers or an allele for purple flowers, the other parent plant providing the second allele required to make up the pair.

The Law of Independent Assortment states that all the pairs of alleles split up separately, so any mix of alleles from the parent is possible in the gamete.

The Law of Dominance states that if an organism has an allele for a dominant trait,

Brno monastery, where Gregor Mendel conducted his experiments into inherited characteristics in plants, now houses a museum in his name.

then this is the trait that will be expressed.

The Mendelian laws of inheritance don't fully explain how traits are inherited – not all alleles are strictly either/or, with a dominant and recessive form. Some traits have more than two alleles, and some are the result of alleles acting together. But Mendel's account was a good enough statement of the case for genetics to have got off to a fair start.

Sadly, that didn't happen. Mendel presented his research in 1856 and published it in 1866, but its reception was largely as a study in hybridization, so it did not attract the scientific audience it deserved. It consequently had no impact on the development of evolutionary thought until the 20th century. Mendel himself gave up most of his research activities in 1868 when he was appointed abbot of the monastery.

Looking at cells

Mendel could propose no mechanism at the cellular level for what he observed in his generations of pea plants. Indeed, at the time of his work, no one was aware of how cells divided in a growing organism or how the germ cells are produced. These two different types of cell division, mitosis and meiosis respectively, were revealed in the late 19th century.

The material of inheritance had already been seen, though not identified, before Mendel began experimenting in the garden and laboratory of the monastery. In 1842, the Swiss botanist Karl von Nägeli saw a tangled network of stringy structures within the cell's nucleus, but did not recognize them as discrete chromosomes. He assumed that the web he saw formed a framework

spread throughout the whole organism.

Forty years later, the behaviour of chromosomes during cell division became apparent. In 1883, Belgian zoologist Edouard van Beneden observed meiosis working on the eggs of *Ascaris*, a roundworm parasite of horses. He saw that

Mendel's laws of inheritance show the transmission of characteristics from one pea generation to the next.

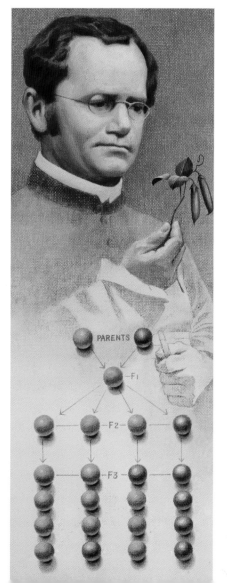

Heinrich von Waldeyer-Hartz gave chromosomes their name.

the number of chromosomes (though he did not use the term) in the germ cells is half the number in normal somatic (body) cells.

Mitosis was studied, and named, by the German biologist Walther Flemming in 1882. He observed the separation and then duplication of the chromosomes before the cell divides, each new cell having the full number.

Chromosomes untangled

Chromosomes were finally named in 1888 by the German anatomist Heinrich von Waldeyer-Hartz. The name refers to 'chromatin', the mix of DNA and proteins of which chromosomes are made, and the name chromatin in turn reflects the fact that it takes a stain and so appears coloured in preparations for microscopy. In fact, the material of chromosomes had already been discovered in 1869 and given the name 'nuclein', but Flemming was not certain that chromatin and nuclein were the same thing, so used a different name.

Nuclein's role suspected

In 1875, the German zoologist Oscar Hertwig discovered that new life begins with fertilization of an egg cell by a sperm cell; Hertwig concluded that fertilization is a physical and chemical process, and he rejected previous ideas of some kind of essence, spirit or ferment. In 1885, he suggested that nuclein was responsible for both fertilization and the heredity of characteristics.

Putting the pieces together

It took the German theoretical scientist August Weismann (*see* page 137) to work out what was going on within the cell in terms of heredity. Weismann developed an eye disorder which prevented him from carrying on microscopic work relatively early in his career. A fortunate consequence for the history of science was that he spent the following years thinking carefully about the puzzles of heredity and evolution.

Weismann was a Darwinian (he has even been called more Darwinian than Darwin himself), but he recognized early on that the changes wrought by evolution are the exception rather than the rule in terms of heredity. First, it is necessary to establish a stable line of descent between generations; only then can the mechanism of change in that line of descent be examined.

SPLIT AND SPLIT AGAIN

Meiosis is the division of diploid cells (those with a full set of chromosomes) to produce haploid cells (those with half the full complement of chromosomes). It produces the germ cells, or gametes – eggs and sperm.

Mitosis is the division of diploid cells to produce further diploid cells. It is the process by which new body cells (somatic cells) are produced for tissue growth and repair.

Weismann published his ideas in 1885. He made a very clear distinction between the germ cells, which (in combination) could give rise to a new organism, and the somatic cells which form the bulk of the body. He claimed that genetic information is transferred in only one direction: the germ cells combine to produce somatic cells, passing on genetic information to them, but the production of new germ cells is isolated from most of the body and cannot incorporate new information from it. Consequently, no kind of Lamarckian inheritance of acquired characteristics is possible (*see* page 137).

Further, Weismann predicted that in meiosis, the creation of the germ cells involves halving the number of chromosomes, so fertilization involves bringing together two half-sets to produce one full set of chromosomes, which defines the new organism. This was proven correct in 1888 by the experimental observations of two German biologists, Theodor Boveri and Eduard Strasburger. Weismann's explanation of how two haploid germ cells produce a fertilized, diploid egg with the full set of genetic information needed to produce a new organism is considered one of the most important discoveries in biology.

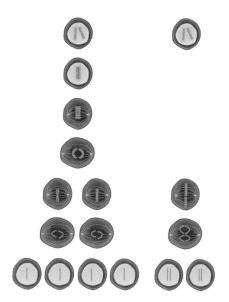

Two forms of cell division – meiosis (left) and mitosis (right).

Mendel rediscovered

Mendel deduced his laws without knowing about the existence of chromosomes and genes, so he could not fully explain the physical process by which traits were passed between generations. When his work was rediscovered, it was possible to make that connection.

In 1900, three European botanists, all working independently of one another on plant hybrids, rediscovered Mendel's work. Dutchman Hugo DeVries, German Carl Correns and Austrian Erich von Tschermak each encountered Mendel's paper when reviewing the literature and each published his own findings with reference to Mendel's work in the same year, more than 40 years after Mendel had

August Weismann proposed that the germs cells can determine the somatic cells in a new organism; the somatic cells cannot change or determine the germ cells.

first made the discovery. DeVries linked heredity with Darwin's theory of evolution, suggesting that mutation was the means by which variant features emerged. Beneficial mutations are favoured by natural selection, and are retained and proliferate, enabling the evolution of a species. At last, the link between heredity and evolution could be explored.

> 'The association of paternal and maternal chromosomes in pairs and their subsequent separation during the reducing division . . . may constitute the physical basis of the Mendelian law of heredity.'
>
> Walter Sutton, 1903

From chromosomes to genes

In 1903, Theodor Boveri and the American geneticist Walter Sutton independently suggested that paired chromosomes are the units of inheritance as described by Mendel. The theory is often known as the Boveri-Sutton chromosome theory.

Until this point, most scientists believed all chromosomes were equivalent to one another. But the Boveri-Sutton hypothesis proposed that the chromosomes differ, and that the splitting and pairing of chromosomes from the male and female parent was the reason for variation between individuals and the means of Mendelian inheritance.

The idea that each chromosome is different and carries particular inherited characteristics raised a new problem, though. There are clearly more characteristics than there are chromosomes, so the mechanism must be more complex than one chromosome per characteristic. Sutton recognized this, but assumed that 'all the traits associated with a given chromosome must be inherited together'.

In 1905, it was demonstrated that some traits in the sweet pea are always inherited together. This linkage between features seemed to support the idea that each chromosome was inherited (or not) in its entirety. But the US geneticist Thomas Morgan would find a different explanation, which led him to produce the first gene map.

Fruit flies to the fore

Around 1908, Morgan and an undergraduate working with him, fellow American Alfred Sturtevant, established a research laboratory at the University of Columbia

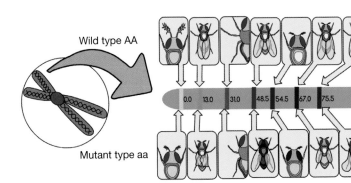

Wild type AA

Mutant type aa

*Part of a gene map of the common fruit fly (*Drosophila melanogaster), *showing the function and relative location of genes.*

that became known as the 'fly room', where they worked intensively on the common fruit fly (*Drosophila melanogaster*). Fruit flies have a brief lifespan and, though tiny, are easy to keep and breed, making them ideal model organisms for genetics research. The fly room became world famous, and an international exchange market in mutant fruit flies grew up, with the fly room at its hub.

GENES AND GENETICS

In 1909, the Danish botanist Wilhelm Johannsen first used the word 'gene' to denote the functional unit of heredity. The term 'genetics' had already been coined in 1905 by the English biologist William Bateson.

Eye, eye, XY

In 1910, Morgan discovered a mutation in his fly jars – a male with white eyes instead of red. Breeding from this single male, he found all the first generation had red eyes, but by then interbreeding from these offspring, some of the next generation had white eyes. He concluded that, following Mendel's prediction, red eyes was a dominant trait and white eyes were recessive.

But there was more. The balance of red and white eyes was not equal between the sexes: if the mother had white eyes, female offspring might have red or white eyes, but male offspring always had white eyes regardless of the colour of the father's eyes. Looking at their chromosomes under the microscope, Morgan discovered that the females had four pairs of chromosomes shaped like an 'X'. The males had three such pairs but the fourth pair of chromosomes in the males had one shaped like an 'X' and one shaped like a 'Y'. Morgan had discovered the XX and XY chromosomes that

determine sex. Further, he worked out that the gene for eye colour must be on the X sex chromosome; whenever the mother supplied a gene for white eyes this was the only eye-colour gene that the male offspring possessed, so it was always expressed. The male passed on a Y chromosome that had no eye-colour gene.

Genes together and apart

Further work in the fly room revealed that some characteristics are often inherited together. Morgan assumed that these must be on the same chromosome. However, they were not always inherited together so separate genes were clearly involved. Morgan proposed the idea of 'crossing over', whereby part of a chromosome is exchanged for the corresponding part of the paired chromosome. He demonstrated that the chances of genes being kept together when a chromosome is chopped up in meiosis depend on the physical proximity of the genes on the chromosome.

How crossing over can lead to a chromosome made up of genes from two chromosomes.

This makes sense: if chromosomes split and recombine fairly randomly, it is more likely that adjacent genes will remain together than those further apart. Genes that are on the same 'chunk' of chromosome are then inherited as a group.

By analyzing how often certain characteristics are inherited together, Morgan calculated the distance between genes on the chromosomes. The result was the first-ever gene map, drawn up by Arthur Sturtevant in 1911.

The ability of genes to move around on chromosomes was demonstrated conclusively by American geneticist Barbara McClintock, working on maize plants in 1931. Her work was met with criticism at the time, and her achievement went unrecognized for many years. But in 1983 she won the Nobel Prize for her work on mobile genetic elements ('jumping genes').

Nailed: DNA and heredity are inseparable

Morgan was awarded the Nobel Prize for his work on genetics, but it was still not clear that the traits he was looking at were coded in DNA. The breakthrough discovery, that DNA carries genetic information, came in 1928. The chemical itself, however, had been on biologists' radar for a while.

From nuclein to chromatin to DNA

In 1869, soon after Mendel published his findings, the Swiss biologist Friedrich Miescher isolated the substance that he called 'nuclein', because he found it in the nucleus of white blood cells. He acquired his samples by washing the pus from used bandages, a process that was probably sufficiently unpleasant to prevent anyone else wanting to muscle in on his research area.

Nuclein was a combination of nucleic acids (RNA and DNA). Nine years later, in 1878, the German biochemist Albrecht Kossel showed that nuclein contained a protein component and a non-protein component. He identified the latter as nucleic acid, and over the period 1885–1901 he found the nucleotide bases which make up DNA and RNA. In DNA, these are adenine, cytosine, guanine and thymine. In RNA thymine is replaced by uracil.

In 1919, American biochemist Phoebus Levene identified the base, sugar and phosphate nucleotide unit and suggested that DNA is a string of nucleotide units linked together through phosphate groups. But Levene concluded that the structure simply

Barbara McClintock's work on maize showed how genes could move around on chromosomes.

The remarkable colours and patterning of the giant atlas moth are all coded into DNA.

repeated nucleotide units in a fixed sequence. This would not give any potential for coding, so his work made it look less likely that DNA was the material of inheritance.

Progress with DNA pretty much halted until the 1940s. Perhaps unsurprisingly, the idea that all the complexity of defining a living organism was incorporated into the arrangement of atoms was difficult to grasp. This was the opposite of the grand view of humans as God's special creatures, which had prevailed just 100 years before. To many, it seemed impossible.

Quick and cheap

In the late 1940s, the Austrian biochemist Erwin Chargaff discovered that the bases in DNA always occur in pairs – a crucial characteristic of the molecular structure that enables it to replicate itself. He claimed later that he only chose to study chemistry for his doctorate because of financial constraints. At a time when students had to pay for their own scientific equipment, chemistry – about which he knew nothing – was the cheapest option. It required, he recalled, 'neither too much time nor too much money'.

Dead mice reveal all

The Canadian-born physician Oswald Avery was working on a discovery, made in 1928, that if mice were injected with a deadly form of pneumonia that had been killed by heating along with a live but non-deadly form, they sickened and died.

'The supposition that particles of chromatin, indistinguishable from each other and indeed almost homogenous under any known test, can by their material nature confer all the properties of life surpasses the range of even the most convinced materialism.'

William Bateson, biologist, 1916

Investigation after death revealed a live deadly form in their bodies. It was clear that genetic material was being transferred from the heat-destroyed form to the less dangerous form, effectively reincarnating the deadly form.

Avery established in 1944 that DNA is the 'transforming' agent which enables one form of the bacterium to take on properties of the other form. He did this by inactivating DNA, RNA and protein in turn in the heat-destroyed form, and finding that if he destroyed the DNA, the transfer did not take place. He went on to show that just providing the DNA, instead of the full cells, also facilitated the transfer of material. Avery's work was the first to show that bacteria have DNA. At first there was resistance to his findings, as many people wanted the transforming agent to be protein.

In 1952, two American researchers, the bacteriologist Alfred Hershey and geneticist Martha Chase, demonstrated even more clearly that DNA is the means of genetic transfer. They used bacteriophages, tagged with radioactive sulphur or

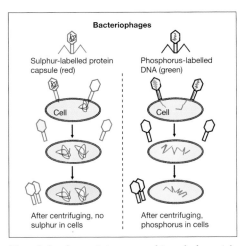

Tagged phosphorous is incorporated into the bacteria's DNA, but tagged sulphur is not.

phosphorous, to infect the bacteria *Escherichia coli* (*E. coli*). Sulphur is incorporated into protein, and phosphorous into DNA. The phage infects the bacterium by injecting its own genetic material into the cell, which then hijacks the cell, using it to produce more viral particles. After growing the phages and then separating phage and bacteria, they found that phosphorous, but not sulphur, had been transferred into the bacteria, demonstrating that the protein coat of the phage was of

SEEING WITH X-RAYS

X-rays were discovered accidentally by the physicist Wilhem Röntgen in 1895. In 1912, a fellow German physicist, Max von Laue, developed the technique of X-ray crystallography. By examining the patterns produced when X-rays spread out (diffract) after passing through substances such as biological molecules and crystals, he found he could discover their internal structure. The British physicist William Astbury, who pioneered the use of X-ray crystallography for examining complex biological molecules, first applied the process to DNA in 1937. Rosalind Franklin was extremely proficient at the technique of X-ray diffraction. It was her X-ray photographs of DNA that enabled Francis Crick and James Watson to work out the exact structure of DNA in 1953.

no importance in the transfer mechanism. Although Hershey and Chase did not claim to have shown that DNA is the means of genetic transfer, the structure of DNA was revealed the following year and answered the question conclusively.

DNA revealed

The race was on to discover the molecular structure of DNA. Scientists at universities around the world turned to the problem, including the Nobel-prizewinning American biochemist Linus Pauling in California and two young molecular biologists in Cambridge, England. They were the Englishman Francis Crick (1916–2004) and the American James Watson (b.1928). When Pauling published his solution, the younger biologists feared he had beaten them to it – but he proposed a triple spiral that would not work in the way DNA works. Aware that as soon as his model was criticized Pauling would rush to fix it, Crick and Watson realized they would have to redouble their efforts if they were to beat him. They succeeded, becoming perhaps the most famous names in 20th century biology. In revealing the structure of the DNA molecule, they paved the way for all future work in genetics and heredity.

But Crick and Watson formed only half the team that did the pioneering work. The other two members were New Zealand-born molecular biologist Maurice Wilkins and the British chemist and X-ray crystallographer Rosalind Franklin; unlike Crick and Watson, these two did not get on well. Wilkins thought of using X-ray crystallography (*see* box opposite) to try to discover the molecular structure of nucleic

acid. (Watson had heard Wilkins talk in Naples, Italy, and was won over by the idea of joining the race to understand DNA.) At the same time, other research teams were using X-ray crystallography to investigate the structures of other important biological molecules – haemoglobin (which carries oxygen in the blood) and myoglobin (which stores oxygen in the muscles).

At Wilkins' suggestion, and without Franklin's permission or even knowledge, Crick and Watson studied a very high-quality X-ray photograph of DNA that Franklin had prepared. This was far better than the images available to Pauling. The X-ray diffraction image, now known as 'Photograph 51', clearly showed that DNA has a double spiral, or helical structure. This, together with previous knowledge

Watson (left) and Crick with their model of the molecular structure of DNA.

about how the bases paired up, gave Watson and Crick the final pieces of the puzzle. In 1953, they described the structure of the DNA molecule: a double helix, with the two outer strands linked by rungs comprising paired nucleotide bases. The bases are held together by hydrogen bonds – an electrostatic attraction between a hydrogen atom in one molecule and a highly charged atom such as oxygen or nitrogen in another. Crick and Watson famously rushed into the Eagle pub opposite their laboratory to announce their discovery. All four scientists also followed the more regular route of publishing papers in *Nature* (in the same issue) later that year.

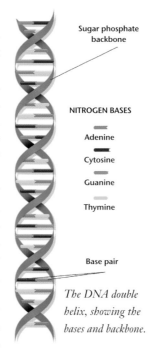

Sugar phosphate backbone

NITROGEN BASES

Adenine

Cytosine

Guanine

Thymine

Base pair

The DNA double helix, showing the bases and backbone.

With the structure of DNA revealed, the method by which it replicates fell into place. Crick and Watson did not describe this in their original paper, opting instead for the cautious: 'It has not escaped our notice that the specific pairing we have postulated immediately suggests a possible copying mechanism for the genetic material.' It was enough to establish priority, but not enough to put themselves on the line if wrong.

They described how replication works in a paper later the same year. Because the bases always occur in the same pairs, the DNA molecule can easily 'unzip' down the middle, forming two single strands with bases sticking out of the side. Each unzipped strand has the recipe for its partner strand – all the cell has to do to rebuild the entire DNA strand is to add the appropriate paired base to each lone base and finish off with another sugar/phosphate supporting strand. The DNA molecule can reproduce itself, and all the genetic information it holds, by a straightforward process of chemical synthesis. The number of permutations of the bases in a long molecule is sufficient to carry a unique genetic code: 'It therefore seems likely that the precise sequence of bases is the code which carries the genetical information,' Crick and Watson concluded.

THE HYDROGEN BOND SAVES THE DAY

The reason the base pairs are always guanine-cytosine and adenine-thymine is that there are two hydrogen bonds between adenine and thymine but three between guanine and cytosine – it's impossible to build the molecule differently.

The central dogma

Watson taped a note above his desk:

DNA → RNA → protein

It was to remind him not of any chemical transformation but of the flow of information he believed took place. It was to become what he called the 'central dogma' of molecular biology, first stated in 1956. The idea is that RNA plays the part of messenger, reading information from DNA, and organizes the building of a protein. The precise mechanism, which emerged over the ensuing decades, is too complex to go into here. What is most important is that Crick and Watson demonstrated how the 'one gene, one enzyme' model of the action of DNA could work.

The idea that each gene codes for a single protein, such as an enzyme, was first proposed by two American geneticists, George Beadle and Edward Tatum, in 1941, well before the biochemical mechanism could be understood. Their work on red bread mould, *Neurospora*, led them to the discovery that producing a mutant with only one gene differing from the parent was sufficient to prevent the production of an enzyme that the mould needed for metabolism. The final proof that one gene does indeed code for one protein came in 1964.

How to make a protein

It's one thing to know that DNA carries the recipes for proteins, and another to be able to read and follow the recipes.

Proteins are all made up of amino acids. It was clear that the sequence of bases must identify the amino acids needed to make a protein. As there are 20 amino acids and four bases, a sequence of

Marshall Nirenberg uncovered precisely how RNA and DNA code for proteins.

two would not be enough as this gives a maximum of $4 \times 4 = 16$ possible codes. There must, then, be at least three bases to a code, giving a possible $4 \times 4 \times 4 = 64$ combinations. In 1961, American biologist Marshall Nirenberg found the first code for a protein. In an experiment that used artificial RNA to build proteins, he found that a chain of the base uracil (found in RNA) led to the synthesis of a protein made of repeated phenylalaline units. This was proof of the concept that sequences of bases code for proteins: the code existed and could be cracked. The unit of three bases that maps to an amino acid is called a codon. Finding the codons for each of the 20 amino acids would unlock the recipe for making life.

The dark side of genetics: eugenics

Understanding genetics has huge benefits for humanity – but it also opens the door to some very disturbing possibilities. One of those is eugenics – the deliberate manipulation of a breeding population to engineer the genetic make-up of future generations. Early advocates of eugenics included Francis Galton, who wrote in 1904 that 'the aim of eugenics is to represent each class or sect by its best specimens'. He did not want to eradicate differences, he said, but to produce the best 'specimens' of diverse races and types. It sounds relatively benign until you get to sorting out how to

A lecture by Norwegian eugenicist Jon Alfred Mjoen in the early part of the 20ᵗʰ century.

define 'best': we would become 'less foolish, less frivolous, less excitable, and politically more provident than now' and, even more alarmingly, 'we should be better fitted to fulfil our vast imperial opportunities'. His plan was to restrict marriage (and therefore breeding) to move humanity, or at least the British, towards a higher calibre of person.

Galton was by no means the last to entertain such ideas. Adolf Hitler was famously in favour of eugenics, adopting a more brutal approach to removing the people he considered 'undesirable' or 'lesser', including Jews, Romany, homosexuals, disabled people and the mentally ill. But eugenics was dictating policy in the United States long before the emergence of the Nazis. In 1907, Indiana became the first state to enforce sterilization on patients in mental hospitals. In 1907–63, 64,000 Americans (mostly women) were forcibly sterilized through eugenics programmes to 'protect' white 'racial health'. At the same time, educated middle-class white women were encouraged to have more children to reinforce 'good' qualities in the American gene pool. Black and American Indian women with many children, living on welfare, were threatened with withdrawal of welfare if they did not agree to be sterilized, and this continued right up until the 1970s.

A proposal in 1911 to use euthanasia (by gas chamber) to cleanse the American population of undesirable genetic traits was considered too contentious. Instead, other methods were used. In Illinois, people admitted to a mental hospital in Lincoln were fed milk infected with tuberculosis, with a consequent 30–40 per cent mortality. The reasoning went that genetically fit individuals would defeat the infection.

Even James Watson has spoken in favour of using the knowledge that DNA has brought to give evolution a helping hand in turbocharging the human race by screening for genetic birth defects and aborting affected foetuses, and by discouraging carriers of genetic conditions from having children in a drive to 'have better babies'.

> *'Decisions concerning the application of this knowledge must ultimately be made by society, and only an informed society can make such decisions wisely.'*
> Marshall Nirenberg, 1967

Mapping the genome

Once it became apparent that each gene comprises a series of codons, each of which specifies a single amino acid, it was also clear that it would be possible – at least in theory – to make a very long list of all the codons and work out what each gene does. It's not necessarily the case that knowing which protein is synthesized means we know what the protein does in the organism. But the first step must be to identify the sequence. The task of listing all the genes of an organism – its genome – is known as mapping. The first genome ever to be mapped was of an RNA virus in 1975, Bacteriophage MS2. The first DNA genome sequenced was of Phage -X174. The British biochemist and pioneer geneticist Fred Sanger sequenced its 5,386 base pairs in 1976. Sanger introduced a new and much faster method of sequencing DNA in 1977, winning his second Nobel prize for the work.

Other genomes followed thick and fast, with the first draft of the human genome completed in 2003. Individual genomes can now be sequenced relatively quickly and cheaply: James Watson's genome was fully mapped in 2007, taking two months and costing less than US$1 million. By 2016, the price had dropped to around $1,000.

The genome provides immense potential not just for genetic medicine, but for understanding the evolution of humans in relation to other organisms.

Evolution and genetics come together

Weismann was clear that we need to understand the normal workings of heredity before we can understand evolution. The real question was how changes come about and what makes them perpetuate.

Speciation

The example of Galapagos finches showed how separate populations of the same species that had become isolated could evolve along different paths as they adapted to different conditions. Darwin saw evolution as a steady, gradual process. One logical consequence was that at some point there would be a grey area where we would be unable to say whether both populations were still one species or whether they had become two.

Fred Sanger developed rapid gene-sequencing methods.

'[The draft genome] is a history book – a narrative of the journey of our species through time. It's a shop manual, with an incredibly detailed blueprint for building every human cell. And it's a transformative textbook of medicine, with insights that will give health care providers immense new powers to treat, prevent and cure disease.'

Francis Collins, the director of the National Human Genome Research Institute (2001)

171

The problem of speciation (the emergence of new species), and of defining what a species is, resulted directly from Darwin's work, producing a crisis of confidence in biologists. The question of whether a species is anything besides a category in the mind of the observer strays into philosophical territory. Certainly the old idea that species were pretty much fixed had now to be abandoned, and the point of transition or emergence raised new questions about the concept of 'species'.

Theodosius Dobzhansky, Russian geneticist.

Towards a synthesis

Two scientists working in the mid-20th century would produce from this dilemma the marriage of evolutionary theory and genetics that is still the principal paradigm in evolutionary biology. One was Theodosius Dobzhansky, born in the Ukraine (then part of the Russian Empire). The other was the German Ernst Mayr.

Dobzhansky moved to the USA in 1927 at the age of 27 where he worked first with Thomas Morgan in the fly room at Columbia University. In 1926, the American geneticist Herman Muller (1890–1967) had discovered that radiation could be used to increase the rate of mutation in fruit flies, making the study of mutations and heredity easier. In 1937 Dobzhansky published *Genetics and the Origin of Species*, one of the most important texts on evolutionary genetics. In it he defined evolution as 'a change in the frequency of an allele within a gene pool', and promoted the idea that natural selection takes place through mutation in a gene pool, with selection favouring beneficial mutations.

As Muller would explain it later, mutations are random – they aren't directed towards any end. Many are harmful or even lethal, but a few will be beneficial to the organism. Beneficial mutations will be perpetuated, as the organisms with them will

GENETIC DRIFT

Genetic drift occurs when, by chance, the prevalence of some alleles over others is established in a population. The concept stems from the work of American geneticist Sewall Wright, published in 1929, though Wright saw it as a directed process of change. Drift is most common when a small population is separated from a larger population. Over time, some alleles disappear completely, and others become universal, changing the features of the population. It tends to happen with alleles that do not present any particular advantage or disadvantage. So, for example, animals living in an area with no predators might lose their defensive camouflage colouring because it was not being reinforced by natural selection.

be successful and breed, passing them on. In fact, most mutations have neither a positive nor negative impact on the functioning of the organism and can contribute to evolution by 'genetic drift' (*see* box opposite). These two impulses lie behind the way organisms change over time.

Dobzhansky's work influenced Ernst Mayr, who was initially an ornithologist. Mayr was interested in speciation – and the point at which species diverge – but also in the whole problem of the definition of species. In *Systematics and the Origin of Species* (1942) he discussed several 'species concepts' – ways of defining species. He favoured the biological species concept, which defines a species as a group that can interbreed, and not breed with others. (This cannot be the only definition of species because it does not take account of organisms that reproduce asexually.) More species concepts have been developed since Mayr's time, one of the more influential (and recent) being the phylogenetic concept. This defines a species as the smallest group that can be distinguished by a unique set of genetically determined traits. It involves looking at the DNA of a potential species and comparing it with related species.

Drifts and rings

When a population within a species-group becomes isolated it can change through both genetic drift and natural selection to the point where it becomes a distinct species, no longer capable of breeding with the rest of the original group. This is called allopatric speciation. Mayr also identified a subset of allopatric speciation, peripatric speciation, which occurs when a small population on the edge of the range of a large population develops separately. An example is the evolution of the polar bear. At an extreme edge of the range of the brown bear, bears were subject to different evolutionary pressures from the majority of the widespread population and evolved differently. (Brown bears and polar bears can still interbreed.)

Polar bears and brown bears could be in the process of developing into separate species, or climate change might bring them back together.

Mayr also recognized that species and subspecies do not really represent a problem for biologists – they just show evolution in action. Variation within a species (subspecies), such as birds with long or short tails, might arise in different areas of the birds' range while the population can still interbreed. The variation might ultimately lead to different species – or it might not.

In *Systematics*, Mayr also gave an account of ring species. These occur when minor variations in a species occur in adjoining parts of the range, forming a ring (geographically). An example would be a population that circles a mountain. Almost all the way around the ring, adjacent groups can breed with one another, but at one point the differences between the starting population and the final subspecies have become so great that the adjacent variants can no longer breed.

Stephen Jay Gould proposed that evolution proceeds by leaps and bounds – and long, lazy periods.

Leaps and bounds

Darwin had been adamant that evolution is a slow process which takes place through many small incremental changes over time. This view predominated until the 1970s. Then, in 1972, two American biologists, Stephen Jay Gould and Niles Eldredge, proposed a model of 'punctuated equilibrium'. This suggests that there might be long periods of little or no change, and then a flurry of evolutionary activity. The idea was based on Mayr's work and their own observations as paleontologists. They had found that periods of stasis dominate in the fossil record, and when there is change it is generally rapid. They proposed that rapid change generally happens in an isolated group or at the edge of a population's range, since accumulated changes are more likely to survive in a small breeding population. Because so few organisms are ever fossilized, the stage of rapid change is rarely represented in the fossil record.

Evidence of evolution following the pattern of punctuated equilibrium has been found in some species, including some types of bryozoan (a coral-like sea creature). The relative importance of gradual and rapid evolutionary change is still debated and researched.

Eat your ancestor

The pattern we generally think of when contemplating evolution is the change from one type of organism to another – from a dinosaur to a modern bird, perhaps, or a fish to an amphibian. But an altogether different type of step was proposed in 1910 by the Russian botanist Konstantin Mereschowsky.

Chloroplasts are the organelles in plant cells which enable green plants to

Mereschkowsky suggested that one organism could entirely incorporate another.

photosynthesize. In 1883 French botanist Andreas Schimper noted a similarity between chloroplasts and cyanobacteria. Mereschowsky took this further, suggesting that chloroplasts are the evolutionary descendants of something like cyanobacteria that have been absorbed and incorporated into green plants. He suggested that at some point in evolutionary history, some unicellular organisms that existed symbiotically inside others became wholly integrated with them, turning into a single organism. In this model, plants contained single-celled organisms like cyanobacteria which carried out photosynthesis, benefiting the plant. The incorporated microorganism became an organelle in the cells of the host.

In the 1920s, the American biologist Ivan Wallin extended the idea to mitochondria, the organelles that are often called the 'power house' of the cell as they generate energy from carbohydrates. Again, he suggested that they were previously independent organisms which had become co-opted into the cell. Although the theory was largely ignored in the early 20th century, it came to prominence again in 1967 with the work of another American, the evolutionary biologist Lynn Margulis. If the theory is correct, mitochondria would have started out as prokaryotic organisms

living in symbiosis with their host cells. During the process of evolution, they became an integral part of it, a single eukaryotic organism. This would be an example of a large leap as predicted by Gould and Eldredge, rather than a gradual change through small adjustments. The DNA of mitochondria is separate from the DNA of the chromosomes in the cell nucleus. The existence of this separate genome supports the theory that the mitochondria were originally separate organisms.

> '*Evolution is a fact. For the evidence in favor of it is as voluminous, diverse, and convincing as in the case of any other well-established fact of science concerning the existence of things that cannot be directly seen, such as atoms, neutrons, or solar gravitation.*'
> Hermann Muller, US geneticist and Nobel Laureate, 1959

The green algae spirogyra contains spiral-shaped chloroplasts that were once independent organisms.

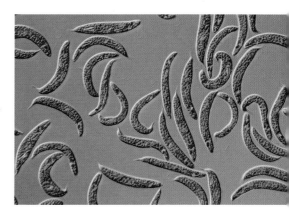

CHAPTER 8

We're all in this **TOGETHER**

'Anything which is contrary to nature is dangerous.'

Theophrastus, 4th century BC

Organisms don't exist in isolation. We are all part of an interacting community of living things within a local environment (ecosystem), part of a much larger habitat of plants and animals (biome) and, ultimately, of all the living things on the planet (the biosphere). How plants and animals interact in this complex web has been one of the most significant discoveries of the 20th century, yet it is arguably also the oldest of insights and instincts.

Monarch butterflies migrate annually from Canada to Mexico, but climate change is damaging the Mexican forests where they overwinter, and threatening their survival.

All one

The idea that the natural world is a single whole is both very old and very new. Ancient Chinese and Indian belief systems are rooted in the notion that everything is part of a great cycle and all living things are linked together. Although originally built into a spiritual framework, this early awareness of the myriad connections between living things is reflected in modern notions of food webs, chemical cycles, symbiotic relationships and fragile ecosystems. But between early notions of spiritual unity and modern theories of finely balanced and fragile connections lie 2,000 years of bafflement, blindness and, finally, discovery.

From gods to natural causes

Many early civilizations first saw the natural world as the stage of the gods. To the Ancient Greeks, Poseidon's thundering horses were behind earthquakes and the angry Zeus could send a flood. Viewing events as divinely orchestrated did little to help humankind understand the environmental impact of their activity. The gods obscured natural cause and effect. If digging a canal caused a flood or a drought, for example, it was a sign of divine anger at the canal rather than evidence that the area could not sustain such a construction. Second guessing what the gods wanted wasn't the best way to approach sustainable management of

This carving of the 'Wheel of Life' at Dazu, China, shows the realms into which a soul can be reborn, linking all life forms.

natural resources. From the 4th century BC, a more rational approach to the interaction of humans and the environment emerged in Greece.

> 'What now remains compared with what then existed is like the skeleton of a sick man, all the soft and fat earth having wasted away, and only the bare framework of the land being left.'
>
> Plato, *Critias*, 4th century BC

Hippocrates wrote an account of environmental determinism, suggesting the types of illness which affected people depended on where they lived. A person who lived in a cold, damp place, for example, would most likely fall prey to diseases with a cold, damp character. This suggests a one-way conduit, with the environment affecting the organism (humankind, in this case); there is no suggestion that the organism might in turn affect its environment. In

In Ancient Greek mythology, the natural world was a stage on which the gods acted out their dramas.

Sophocles' play *Antigone*, the chorus describes humankind's impact on nature, but this is presented as harmonious rather than intrusive.

Sophocles' account is a display of humanity's dominion over nature, most particularly over individual animals. This is the legacy of the Neolithic revolution that brought farming, the start of humankind's disproportionate influence on the rest of the natural world.

> 'The thoughtless tribe of birds,
> The beasts that roam the fields,
> The finny brood of ocean's depths,
> He takes them all in nets of knotted mesh,
> Man, wonderful in skill.
> And by his arts he holds in sway
> The wild beasts on the mountain's height;
> And brings the neck-encircling yoke
> On horse with shaggy mane,
> Or bull that walks untamed upon the hills.
>
> Sophocles, *Antigone*, 441 BC

A more nuanced account of how people had affected the entire environment is given in Plato's *Critias*. Although couched as a description of Atlantis, it really describes the area around Athens as changed by the Athenians. Plato remarks on deforestation, soil erosion and the drying up of streams – a familiar story.

Two-way traffic

Theophrastus, famous for his work on botany, was the first to write from a truly ecological perspective. His accounts of plants place them in relation to their environment – to the climate and aspect of the land where they are found – and how they are adapted to prevailing conditions. So a particular type of plant might prefer shady areas to full sun, sandy soil to heavy soil, growing on a slope or in a damp area, and will not thrive if planted elsewhere. He often talks about a specific example – a particular tree that he locates so that contemporary readers would have been able to identify it, or one of the botanical specimens that belonged to Alexander the Great.

The olive tree needs a hot, sunny aspect and can survive drought.

Theophrastus was aware, too, of the relationships between different types of plants. Some grow well together, but others affect their neighbours, as the cabbage harms the vine. He mentioned examples of parasitism and symbiosis, and noticed also that there can be beneficial interaction between animals and plants, as when a jay buries acorns and so helps to propagate new oak trees. He knew that legumes enrich the soil, benefiting other plants, and that rotting leaves help to nourish young seedlings.

Like Aristotle immediately before him, Theophrastus believed that everything in the natural world happened for a purpose and followed laws which could, at least in theory, be discovered and understood. Unlike Aristotle, though, he did not believe that the purpose or function of other organisms was directed towards humankind. For Aristotle, the nature of a tree would relate to what he saw as its main purpose of providing wood for making boats, say, or bearing fruit. For Theophrastus, the nature of the tree would be best suited to its own purposes: surviving in its environment and growing fruit and seeds to produce the next generation. Its behaviour and habits were bent to these ends, not to serving humankind. Unfortunately, his was not the view that would prevail. It would be more than 2,000 years until ecology would emerge from the shadow of Aristotle's utilitarian position on the function of nature.

Seeing connections

Although some of the observations that Theophrastus shared might well have been common knowledge among farmers and gardeners, there seems to have been little intellectual awareness in the West of the interaction between organisms. In the Far East, a very different approach prevailed. The Chinese scientist Shen Kuo (1031–95) saw the use of predatory insects as a means of protecting crops. He was also concerned that the use of pine trees as fuel by the iron industry, and the use of pine soot to make ink, could lead to deforestation. He recommended the use of petroleum instead, which he thought was produced inexhaustibly within the earth. The printing ink he made from the carbon deposited by burning petroleum was more durable than the pine-soot ink it replaced.

The Biblical account of Creation gives humankind dominion over the natural world.

All for one – and we're the one

Aristotle's stance, that the natural world was useful primarily insofar as it serves humankind, accorded well with the Christian position, outlined in Genesis, that all animals and plants are subject to man's dominion. After his works were translated into Latin in the 12th and 13th centuries, the views of Aristotle and the Church reinforced one another, making a formidable barrier of authority that would not easily be challenged.

Physicotheology

It was against this background of intransigent religious doctrine and Classical tradition that proto-ecological thinking emerged in the West in the 17th and 18th centuries. More properly called physicotheology, this new approach to nature led eventually to ecology – but by a roundabout route.

Physicotheology saw the intricate and ingenious detail inherent in nature as the manifestation of divine genius. It inspired the close study of organisms and natural systems as a way of gaining insight into and appreciating divine Creation. It led to an accumulation of

Shen Kuo was a pioneer in looking at the environmental impact of industrial processes.

biological knowledge, further facilitated by the development of the microscope and the scientific method, and expansion of travel and exploration.

At the same time, the perception of the whole of nature as the working-out of God's great and glorious plan promoted ideas of unity in multiplicity, harmony and equilibrium, even a great, smoothly running, semi-mechanical system. It might not sound as though it would yield ecological thinking, but the created world emerged as a seamless system in which every part has its function and all work together (albeit to demonstrate the glory of God). The next step would also be taken in the shadow of the Creationist inheritance, but this time prevailing beliefs would hinder rather than aid progress.

The beginnings of biogeography

When Ferdinand Magellan set sail from Spain in 1522 on the first circumnavigation of the world, the Italian scholar Antonio Pigafetta was on board. During the three-year journey, he documented the fauna and flora of the places Magellan's ships visited. (Pigafetta was one of only 18 of the original 240 men to survive the trip.) One of the things he noted was the contrast between the plants and animals living in the Philippines and those living in the Spice Islands (now the Maluku Islands). He had no explanation for the differences. It was a phenomenon that would be recognized again and again as European adventurers explored the world.

Shifting sands and lands

In 1610, Sir Francis Bacon noted that the West African coastline fitted quite well against the coastline of North and South America. It looked as though the two could be pieces of puzzle pulled apart and separated by the Atlantic Ocean. But in a world believed to have been created in its current state by a God who doesn't change his mind about the positioning of an ocean, it remained a curious observation and nothing more.

Many more people were to notice this similarity over the years, but until the 20th century no one had a plausible explanation for it.

Here and there

Linnaeus, and other plant and animal collectors in the 18th century, noticed that organisms are not evenly spread around the world, and that different types of land and climate produce different types of plants and animals. Further, there are

Millions of years ago, the continental landmasses were crowded together in a single supercontinent.

PERMIAN PERIOD

250 MILLION YEARS AGO

some similarities between those that inhabit similar environments, even if they are separated by thousands of miles.

Linnaeus, who was devoutly Christian, explained this in terms of the distribution of animals leaving Noah's ark after the flood. His 'Paradisical mountain' hypothesis states that at the end of the flood, the ark came to rest on a great mountain near the equator. All the animals of the ark spread out to their appropriate niches on the mountain, determined by elevation. As the flood receded, the continents expanded and the animals migrated to new sites. It was a neat, if incorrect, solution that did not challenge the Biblical narrative.

Others took issue with the distribution of animals. English polymath Thomas Browne (1605–82) remarked that 'How America abounded with Beasts of prey and noxious Animals, yet contained not in it that necessary Creature, a Horse, is very strange.'

> '*Nature is God's law, placed in all things during creation, according to which they multiply, sustain, and destroy themselves.*'
> Carl Linnaeus, *Politiae naturae*, 1760

The proto-evolutionary theory of Georges-Louis Leclerc, Comte de Buffon, took a similar line (*see* page 133). He claimed that organisms emanated from a single central point at the North Pole and developed according to the environments in which they found themselves, with those living in similar environments acquiring similar adaptations, even if they were widely separated geographically. This became

Alexander Humboldt.

known as Buffon's law, and was the principle of biogeography.

Leclerc criticized Linnaeus's theory on two counts: different regions with the same climate had similar species, and if animals were not capable of adaptation, as Linnaeus claimed, they would not be able to cross some of the more hostile environments to get to the places where they are now (or were then) found.

In both Linnaeus's version and Leclerc's account, the onus to move was on the animals, and not on the land they inhabited.

Specimens in context

The biogeographers of the 19th century worked on the geographical distribution of plants and animals, relating organisms to the environments in which they were found. Perhaps the most important of the biogeographers was the German naturalist and explorer Alexander Humboldt (1769–1859). On a long expedition to South America, he collected and described many new species. Unlike the cataloguing collectors of the previous centuries, though,

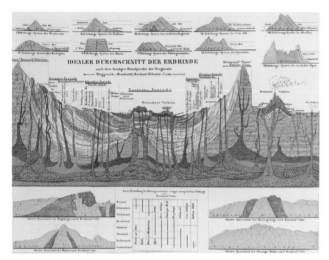

A map from the atlas that accompanied Humboldt's Kosmos.

(1805), laid the foundations for biogeography. He was the first to describe the increasing biodiversity in the tropics.

Humboldt's work was not restricted to botanical discoveries; in addition, he studied physical geography, volcanoes, geology and mineralogy. His wide range of interests underpinned what became his life's ambition – to unify the sciences in a comprehensive account of the whole of nature. The result was his five-volume *Kosmos*, published over the period 1845–62 (the last volume after his death, and unfinished). He intended it to be a compendium of the entire global environment – a stunning first work in the emerging discipline of ecology.

Humboldt did not study his specimens in isolation. A plant specialist, he took a new approach, trying to discover and record the relationship between plant species and the environments in which they lived. He related plants to the climates in which they were found and defined vegetal zones by reference to latitude and altitude. His most famous work, *Idea for a Plant Geography*

The moving Earth

Inevitably, biogeography revealed again the patterns in the distribution of plants and animals that had puzzled earlier explorers and collectors. This time, though, the patterns were tracked and investigated more carefully. The new geological theories of Lyell and others (*see* pages 141–2) had cast doubt on the literal interpretation of the Creation story, and the immutability of the Earth no longer went uncontested.

In 1858, the English lawyer and ornithologist Philip Sclater established the six zoological regions (or ecozones) of the Earth, naming them the Palaearctic,

During his visit to South America, Humboldt learned a few words of the language of a recently extinct tribe, the Aturès, from a parrot.

Aethiopian, Indian, Australasian, Nearctic and Neotropical. These zoological regions are distinct.

Sclater proposed that a lost continent, which he called 'Lemuria', had once joined India and Madagascar but now lay beneath the Indian Ocean. This would, he felt, explain the presence of lemurs in both countries, but their absence from Africa (which is much closer to Madagascar than is India). This was a step in the right direction as it acknowledged that the geography of the world might not be fixed for all time.

Walking over the sea

In 1845 the differences which Pigafetta had noticed between the animals of the Philippines and the Spice Islands were noted again by the British navigator George Windsor Earl. He remarked that the islands on the west were separated from Asia by a shallow sea and had animals typical of the Asian mainland, while the islands to the east had marsupials, similar to those in Australia. Alfred Russell Wallace, co-discoverer of evolution, also studied the area. He defined a line, now known as Wallace's Line, which runs between Borneo and Sulawesi and between Bali and Lombok, with the Asian islands to the west and the Australasian islands to the east. Wallace proposed that

Sclater's ecozones divided the world into regions along zoological lines.

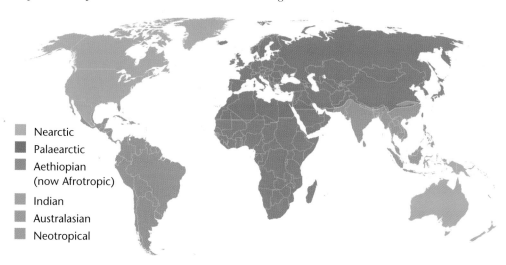

- Nearctic
- Palaearctic
- Aethiopian (now Afrotropic)
- Indian
- Australasian
- Neotropical

the islands to the west of the line had at some point in the past been joined to the landmass of Asia and those to the west had been joined to Australasia. This suggested that animals might have walked across the land, then later have become stranded.

The gap between Bali and Lombok is only 22km (13.6 miles) wide. We know now that during ice-age glacial periods, the sea level was 120m (394ft) lower than now. This was sufficient for the islands to be joined to their respective landmasses, as Wallace predicted. The rift between the two continental shelves, though, was too deep to dry out and remained unbridgeable for animals. This final piece of information was not available to Wallace, but now explains the existence of Wallace's Line.

The pieces fit together

At the end of the 19th century, Alfred Wegener, a German meteorologist and geologist, turned again to the congruity of the American and African coastlines. He also examined the rocks on either side of the ocean and the fossils buried within them.

Wegener found identical rock strata in South Africa and southeast Brazil. He pointed also to the presence of fossils of the dinosaur *Mesosaurus* on both continents. *Mesosaurus* could not possibly have crossed the Atlantic Ocean, yet its fossils did not appear to be very different in each place, unlike camels and llamas (which are also related species). He suggested that the continents have moved around, though he could offer no explanation for how this might have happened. This theory became known as continental drift.

His evidence went beyond the match between the Americas and Africa. He noted, too, that coal is present in both Britain and Antarctica, yet coal forms only in hot, wet conditions. Either the climate in these places had changed radically, or the landmasses had moved. Wegener considered it impossible for Antarctica ever to have been warm enough for the formation of coal, so the landmasses must have shifted.

Wegener's idea was largely ignored when he presented it in 1912. But it finally gained acceptance during the 1960s, when

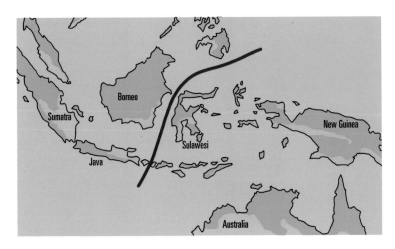

The Wallace Line marks the point at which a land bridge would have existed between Asia and Australasia.

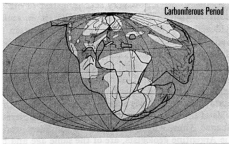

Carboniferous Period

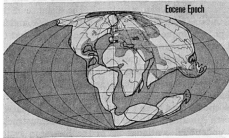

Eocene Epoch

Quarternary Period

Wegener's illustration of continental drift, showing how he supposed the landmasses had moved apart over time.

the mechanism of that movement was uncovered: plate tectonics. Geomagnetic studies of the bottom of the ocean revealed that the seafloor is spreading out from either side of a mid-ocean ridge. It is this process – seafloor spreading – that is pushing Europe and the Americas apart. It became apparent that the thin rocky crust of the Earth sits on top of a slowly moving layer of semi-molten rock called magma. The crust is divided into seven large and several smaller 'plates' which carry the continental landmasses and the oceans. As the magma moves, the plates move. They are pushed apart by magma seeping out at the mid-ocean ridges, and rammed together in slow-motion collisions that force up mountain ranges. The Atlantic Ocean has formed between the plates that carry Africa and the Americas, opening up over a period of millions of years and dividing lands which were once joined together.

For the plant geologists of the 19th century, however, the idea of land moving was inconceivable, but the possibility that land, such as the proposed Lemuria, might have become submerged was entirely reasonable. There are submerged landmasses in many parts of the world, but we know nothing of their previous biology.

Living together

While the plant geologists were interested in finding out which plants lived where, and identifying the adaptations they made to suit their environments, they were largely looking at plants as individual organisms. The move to looking at communities of organisms and their interdependence came during the later 19th century.

Eat me

It is fairly obvious that some animals eat one another and some animals eat plants, so there is clearly some level of connectedness. This much had been noted long ago, and the earliest surviving reference to the notion of a food web is from the Afro-Arab writer al-Jahiz (776–868). But the first attempt to map out the exact nature of the links between organisms that eat one another or are eaten came in 1880.

> 'All animals, in short, cannot exist without food, neither can the hunting animal escape being hunted in his turn. Every weak animal devours those weaker than itself. Strong animals cannot escape being devoured by other animals stronger than they.'
>
> Al-Jahiz, *Kitab al-Hayawan* (*Book of Animals*)

In 1880, Lorenzo Camerano, a 24-year-old assistant in the zoological laboratory in Turin, Italy, published a groundbreaking paper with the title 'On the Equilibrium of Living Things by Means of Reciprocal Destruction'. In it, he proposed two central ideas: that in any community of organisms there is a natural equilibrium level for the population of each type of organism, whether it is plant, herbivore, carnivore or parasite. If this equilibrium is disturbed, the effects will soon be felt in the populations of the other organisms in the interconnected community. He likened this to the

way that any perturbation in the sound wave of an organ pipe propagates along the pipe. Camerano included the first graphical depiction of a food web.

More studies and depictions of food webs followed in the early 20th century, looking at particular environments and ecosystems such as temperate America (1913) and Bear Island (1923), and the food web around herring. The term food 'web' was introduced by the British ecologist Charles Elton (1900–91) in his seminal book *Animal Ecology* (1927). He also introduced the idea of a pyramid of numbers to describe feeding relationships, with the single top predator at the apex of a pyramid that expanded into larger numbers of organisms at lower levels.

Helping hands

Some organisms live together in a close and mutually beneficial relationship rather than one simply feeding on another. This is

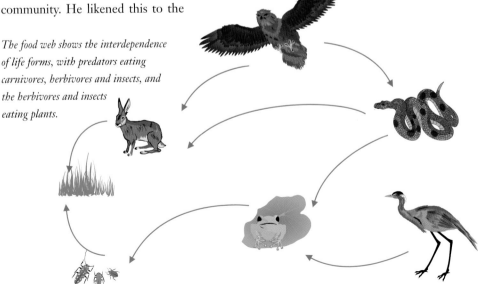

The food web shows the interdependence of life forms, with predators eating carnivores, herbivores and insects, and the herbivores and insects eating plants.

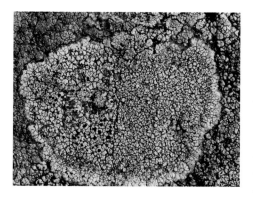

Lichen represents a cosy living arrangement between a fungus and algae.

types interacting and having an effect on one another. In some sense, this realization was already latent in Darwin's *Origin*, with competition between both individuals and species determining the path of evolution within an ecosystem.

Round and round: chemical cycles

It is not just food that passes between organisms in a chain or web. So, too, do chemicals. The German chemist Justus von Liebig suggested in 1840 that plants need a limited number of nutrients, many of which are taken from the soil. When the supply of nutrients has been exhausted by repeated cultivation it must be replenished or crops will no longer flourish. Even Theophrastus had noted that fallen leaves enrich the soil and nourish other plants, and Linnaeus remarked repeatedly on the economy of nature, which recycles everything. He considered nature to be self-cleansing and renewing; which it is, if not overly disrupted. He saw that even apparently discordant events, such as predators killing and eating prey, serve a purpose in controlling populations, thus maintaining harmony and balance in the natural world. Liebig was an early conservationist, encouraging the recycling of sewage, for example.

called symbiosis, and was first described in 1879 by the German surgeon and botanist Heinrich de Bary. The word 'symbiosis' had previously been used of people living in a community; in 1877, the German botanist Albert Frank used it to denote a mutually beneficial relationship among lichens. De Bary defined symbiosis as 'the living together of unlike organisms' – this exactly describes the situation with lichens, which he also studied. Lichen is a composite organism comprising algae (and/or cyanobacteria) growing along the filaments of a fungus. The algae benefits from being protected by the fungus; it anchors itself and draws water and nutrients from the fungal filaments. The fungus benefits because the algae photosynthesizes, producing food for them both.

It was, then, during the final decades of the 19th century and the first decades of the 20th century that ecology really emerged. At that point, biology transformed from being just the science of individual organisms to dealing with whole populations and ecosystems, with organisms of different

> *'Thereby it happens that when animals die, they are transformed into mulch, and mulch into plants, and these plants are eaten by animals, whereby they become parts of animals.'*
>
> Carl Linnaeus, *Journey to Wästgötaland*
> (1747)

So far, so good; but while it was agreed that plants need certain chemicals, it was not clear where these came from. There was disagreement about whether plants take nitrogen and carbon from the air or the soil, for example. Repeating de Saussure's experiments (*see* page 77) more accurately, Liebig demonstrated that there is too little carbon even in humus-rich soil to supply all of a plant's needs, so the plant must take at least some carbon from the atmosphere. He concluded that plants take various nutrients, including phosphorous and potassium, from the soil; carbon and nitrogen from the atmosphere; and hydrogen from the atmosphere and from water. Only legumes such as beans and peas can take nitrogen from the air and 'fix' it by means of microbes living

Justus von Liebig initiated the use of artificial fertilizer, revolutionizing food production.

in nodules on their roots. This mechanism explains why growing legumes in exhausted soil replenishes it, a feature already noted by Theophrastus and exploited by farmers for centuries.

Liebig's work began the age of informed fertilizer use and development in agriculture. Previously, people had noted that using manure as a fertilizer increased crop yield, but the prevailing explanation was that the fertilizer helped to break down humus (mostly dead plant material) making it more readily available for plants to absorb. Liebig made a particularly important contribution in stating that plants can derive nutrients from inorganic as well as organic sources, prompting

THE PARK GRASS EXPERIMENT

The Park Grass experiment is the longest-running study in the world, having been set up by two English agricultural chemists, John Bennet Lawes and Joseph Henry Gilbert, in 1856 to investigate the effects of fertilizer on hay production. The land used, a field of 28,000sq m (33,488sq yd), had been pasture for at least 100 years before the experiment began. Over the course of more than 150 years it has produced unique evidence of the impact of environmental changes on the ecology of a small area, including biodiversity and localized evolution. It has also provided an archive of soil and hay samples that preserve a record of environmental pollution. Some of the plots in the experiment display a variety of meadow flowers that would have been common in the 19th century, but have long since been lost elsewhere. These are the plots that have remained unfertilized, and have 50–60 varieties of plants. The fertilized plots have only two to three species, showing the impact of altered soil pH on biodiversity.

the use of artificial fertilizers. He also popularized the 'Law of the Minimum', first proposed by the German botanist Carl Sprengel (1787–1859). This states that a plant's growth will be limited by the availability of the scarcest nutrient.

Among the studies prompted by Liebig's work was the Park Grass experiment in England to test the impact of fertilizers on the growth of grass (see box opposite).

With von Liebig's work, the chemical activity of the individual plant is placed into the wider context of the plant's place in its environment. The source of nitrogen in the soil, when it does not come from added fertilizer, results from the breakdown of organic matter: the plant is sustained by the other organisms that have lived and died in the area previously, and is tied into a complex ecological system.

A French chemist, Jean-Baptiste Boussingault, discovered that plant growth is limited by and proportional to the amount of nitrogen available, and that adding

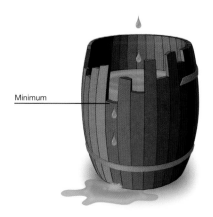

In Liebig's barrel analogy explaining the Law of the Minimum, each stave represents a nutrient or condition needed by an organism. The scarcest resource determines the organism's success, just as the shortest stave determines how much the barrel will hold.

nitrogen and phosphorous at the same time optimizes growth. In a series of experiments reported from 1836 to 1876, Boussingault uncovered most of the nitrogen cycle. The action of microbes on nitrogen in the soil was discovered late in the 19th century, after Boussingault's death.

HABER TO THE RESCUE

As farmers began to use nitrogen-rich fertilizers, the demand for sources rose. At the end of the 19th century, two chemists, Sir William Crookes in Britain and Wilhelm Ostwald in Germany, realized that supplies of guano, manure and nitrogen-bearing minerals were running low and would soon run out. Ostwald was concerned that Germany, with its poor soil, would be vulnerable in any military conflict without supplies of nitrogen for fertilizer and weaponry. The result of the ensuing search for a more secure source led to the German chemist Fritz Haber developing the Haber process for manufacturing ammonia in 1909. Haber's own factory could not work at the scale required and in 1913 he teamed up with the industrialist Carl Bosch, opening the first commercial, large-scale plant to manufacture ammonia for fertilizers. The Haber-Bosch process now provides 50 per cent of the fertilizer used in the world.

Cacti growing on the Uyuni salt flat in Bolivia demonstrate the adaptation of these plants to their harsh environment.

Ecology comes of age

During the 19th century, people first began to consider plants and animals living together as a community. As early as 1825, the French naturalist Adolphe Dureau de la Malle used the term *societé* to refer to plants of different species growing together.

The first true ecologist was Eugen Warming (1841–1924). As professor of plant sciences at the University of Copenhagen, he specialized in plant geography and made frequent trips abroad to study plants in natural environments as diverse as Brazil and Greenland. Warming developed a new approach to botany that focused on how plants adapt to the environment in which they live. He was the first to take account of abiotic factors (non-biological aspects, such as salt concentration, drought and fire) in the distribution and development of species.

Warming was interested in how plants widely separated geographically often seemed to have the same or similar adaptations for dealing with identical problems such as drought, flooding or harsh conditions. Although Warming was strongly influenced by Darwin's work on evolution from the 1870s onwards, there was less connection between evolutionists and ecologists before then. Darwin's work stressed competition and natural selection as the driving factors in evolution, whereas Warming and other ecologists focused on the impact of the environment, and principally its abiotic features, on adaptation.

Warming didn't agree entirely with Darwin's account of evolution, either. He saw such variation within plants of the same species growing in different conditions that he was suspicious of Darwin's model of evolution happening in very tiny steps over extended periods. These variations are now accounted for by phenotypic plasticity – the flexibility within a genotype (genetically distinct type of organism) to adapt its body and behaviour to environmental conditions. Plasticity is more important to plants than to animals as plants can't easily move to a better environment if conditions deteriorate. Warming was a Lamarckian, and believed

DEFINING ECOLOGY

The German naturalist Ernst Haeckel was the first person to use the word 'ecology' (or, rather, *Ökologie*, as it was in German), in 1866. Ecology is a term used rather loosely in the popular press, with 'ecologist' often equated with 'environmentalist'. Ecology is the study of the interaction and relationship between organisms and their environment. An ecosystem is a defined environment and the organisms within it that interact with each other. Ecosystems range from the tiny (a single leaf and its insects and microbes, for instance) to the huge, such as a tropical rainforest spread over hundreds of square kilometres. The environment is the physical location and conditions in which an organism lives – it can even be inside another organism.

AN ECOLOGICAL OPPORTUNITY

In August 1883, most of the island of Krakatau in Indonesia collapsed in a series of massive volcanic eruptions. The effects were devastating for the entire area – it was one of the most violent and deadly eruptions in human history. Global temperatures cooled by an average of 1.2°C for five years. However, it gave biologists an unprecedented opportunity to watch an ecosystem rebuild itself after total destruction.

The first biologists arrived on the island nine months after the eruption and found a single living thing: a spider. They believed nothing at all had survived the blast. Spiders are easily carried long distances by the wind, so the spider was deemed to be the first colonist. For a whole year nothing was seen growing on Krakatau, but a century later the remains of the island were again blanketed in tropical forest, home to 400 species of vascular plants, thousands of types of insects and other arthropods, more than 30 species of birds, 18 species of land molluscs, 17 species of bats and nine types of reptiles. To recolonize the island, 44km (27 miles) from the nearest land, plants and animals had to fly, swim or be carried by wind, water or other animals, including clinging to or carried in flotsam. Surveys of the flora and fauna of Krakatau were carried out regularly and continue today.

that the adaptations plants made to living in a particular environment were passed on to their offspring.

Like many of his contemporaries, Warming was able to reconcile evolution and Darwin's account of the origins of life with his Christian beliefs. He felt that there was no denial of God's creative power, since however the diversity of living species emerged, whether through a six-day creative binge or millions of years of evolution, the rules of nature and physics had been divinely ordered.

In the early 20th century, ecology developed along two lines. On the one hand, the Swedish botanist Rutger Sernander (1866–1944) adopted purely empirical means (based on observation). On the other, fellow Swedish botanist Henrik Hesselman (1874–1943) combined plant biology with plant geography to carry out experiments on plants in their

Research at the Ludwig-Maximilians University in Germany has shown phenotypic plasticity in several water fleas; they can change their body shapes in response to chemicals (kairomones) given off by predators. The form on the left is a standard shape, while that on the right has developed a long helmet (green) and tail (blue, bottom), which may make it more difficult for invertebrates to catch or eat it.

natural settings. This was an innovative step. Hesselman wanted to study not only the external adaptations of plants to changes in their environment, but also the internal physiology of those adaptations. The latter was something that could be undertaken only in the laboratory. The two camps co-existed, with little rivalry between them.

A good starting point

At the start of the 20th century, then, ecology had emerged as a discipline in its own right. It was concerned with the distribution of organisms; the way they adapt to their environments and to change in those environments; how adaptation manifests in physiological structures and processes; and how organisms in an environment are locked together in a web of relationships forming an ecosystem. The first hint that human activities were having an impact on ecosystems came soon after the recognition that ecosystems exist.

Just as plant geographers studied the distribution of plants in different areas, zoogeographers studied the distribution

---

ECOLOGICAL NICHE

An ecological niche is the set of conditions in which an organism can live and prosper. It is often divided into the fundamental and the realized niche: the former is the set of conditions in which an organism can live, and the latter is the set in which it does actually live. The idea of 'ecological niche' was introduced in 1917 and is central to ecology.

The thylacine, or Tasmanian wolf, was a dog-like marsupial which became extinct in the 20th century.

when it was translated from German into English in 1953. He defined ecology as the study of the way animals live, distinguishing between 'existence ecology', which was the relations of organisms to environmental factors, and 'distribution ecology', which was the capacity of organisms to survive in and colonize different environments.

of animals and their adaptations to identical conditions in different places. The Swedish zoogeographer Sven Ekman (1876–1964) identified 'relicts', which are populations of an organism left behind in an environment when the main population has either died out or evolved independently. The relict population may be cut off by changes such as continental drift, changing sea levels, climate change or even predation or competition over part of the original area.

Ekman was one of the first marine biologists, publishing his major work *Tiergeographie des Meeres* (*Zoogeography of the Seas*) in 1935 and becoming influential

From biosphere to noosphere

The Russian geochemist Vladimir Vernadsky (1863–1945) proposed that the development of life and even of human intelligence are essential features of the Earth's evolution over time. He believed that just as the emergence of life profoundly altered the nature of the planet, changing the chemical composition of its atmosphere, oceans and mineral make-up, so the development of human intelligence would itself make its mark on the planet. He popularized the idea of the biosphere first coined by Eduard Suess (1831–1914), and accounted for the balance of carbon, oxygen and nitrogen in the atmosphere by pointing to the activity of living organisms.

Vernadsky's work was not widely distributed in the West, nor was it particularly popular, as it was seen as rather more visionary than scientific. Yet his recognition in the 1920s that living organisms had helped to shape the Earth on which they live, and would continue to do so, would become one of the foundation stones of both ecology and of the environmental movement. He was the first to look beyond the evolution of individual species or even small groups in a contained ecosystem and consider the entire programme of life on Earth and its development in relation to the geological shaping of the planet.

The living Earth

Vernadsky's approach was reprised to some degree years later in the work of Aldo Leopold and James Lovelock. Leopold (1887–1948) was an American conservationist instrumental in driving wildlife conservation as an end in itself.

MARINE BIOLOGY

Marine biology, or the study of organisms that live in the sea, can be said to have started with Aristotle and his description of a wide range of fish, crustaceans (such as prawns), echinoderms (such as star fish) and molluscs (such as squid). There was little further development before the age of European exploration. The English explorer Captain James Cook (1728–79) collected many specimens on his voyages, but more important by far was Charles Darwin, who collected samples from land and sea and pondered the formation of coral reefs and atolls. The first expedition specifically for the purpose of studying the marine environment was led by the Scottish naturalist Charles Wyville Thomson in 1872–6. The expedition collected thousands of specimens and laid the foundations of modern marine biology. In the 1960s and 1970s the first marine laboratories were established. Advances in technology finally made exploration of the deep sea possible, with scuba equipment, submersibles and eventually robotic vehicles to explore even the deep-sea trenches of the abyss.

'A thing is right when it tends to preserve the integrity, stability, and beauty of the biotic community. It is wrong when it tends otherwise.'

Aldo Leopold, 1949

Aldo Leopold made notes each morning on the dawn chorus he heard at his shack in Sauk County, Wisconsin. They now represent a valuable record of how the species living there have changed.

In his early career in forestry, he was employed to kill bears, mountain lions and wolves in New Mexico, where they preyed on livestock. He soon came to respect the animals and turned towards protection rather than destruction. At the time, the only motivation for conservation was to preserve sufficient number of game animals to ensure successful hunting trips – there was no value seen in protecting the biodiversity of the American wilderness for its own sake. He helped found the Wilderness Society in 1935, an organization that promotes the protection of the wilderness for the benefit of the animals which inhabit it, rather than for any utilitarian human reason. For him,

THE LAND ETHIC

'The land ethic simply enlarges the boundaries of the community to include soils, waters, plants, and animals, or collectively: the land . . .

'[Do we not] already sing our love for and obligation to the land of the free and the home of the brave? Yes, but just what and whom do we love? Certainly not the soil, which we are sending helter-skelter down river. Certainly not the waters, which we assume have no function except to turn turbines, float barges, and carry off sewage. Certainly not the plants, of which we exterminate whole communities without batting an eye. Certainly not the animals, of which we have already extirpated many of the largest and most beautiful species. A land ethic . . . affirm[s] their right to continued existence, and, at least in spots, their continued existence in a natural state. In short, a land ethic changes the role of Homo sapiens from conqueror of the land-community to plain member and citizen of it. It implies respect for his fellow-members, and also respect for the community as such.'

Aldo Leopold, *A Sand County Almanac*, 1949

the Society was an embodiment of a new attitude towards the natural world, one marked by an 'intelligent humility toward Man's place in nature'.

Leopold pioneered an approach that put humankind not at the centre of the natural world in America, but occupying a niche within it and with a duty to respect and conserve other organisms. He signposted an attitude towards nature that, for a while, influenced US environmental policy. He called it the 'land ethic', and it set out human responsibilities that extend beyond our private communities into the whole of nature. Leopold was the first to describe what later became known as 'trophic cascade' (the impact of killing one organism or population in an ecosystem), describing the ecological outcome of killing a wolf on a mountain.

Dawning realizations

Leopold's recognition that human activity was having a detrimental effect on the natural environment came at around the same time as the environment began to bite back in a big way. The Dust Bowl years of the American Midwest were the first serious lesson in how mismanagement of natural resources could backfire to wreak havoc on the human inhabitants of the landscape. Unsuitable farming practices that had stripped grass and trees from the land and opened up vast fields led to devastation when a series of droughts struck. Without the grass to hold the soil in place, the topsoil was literally blown away in huge, lethal dust clouds, leaving the land barren and unusable.

The ensuing changes were utilitarian in intent: to protect the livelihood of farmers and the food security of Americans. They were not to preserve the integrity of the ecosystem, as Leopold would advocate. They were, nevertheless, an early attempt to remedy harm done by human activity and an important point of recognition: we share the planet, and its ecosystems are delicate. We tamper at our peril.

And no birds sing

The next leap forward for what would become the environmental movement was the book *Silent Spring* (1962) by the American marine biologist and pioneer activist Rachel Carson (1907–64). Carson became interested in the effects of pesticides on the ecosystem in the 1940s, and

Rachel Carson suffered personal as well as professional attacks for her views.

researched the topic throughout the 1950s. She found scientific opinion divided, with some saying there were no important effects and others claiming evidence of widespread damage to ecosystems from overuse of the pesticide DDT on crops. Her book documented the devastation wrought in bird populations and the transit of DDT through the food chain. It aroused fury in the agrochemical and farming industries, but led eventually to the suspension of DDT as an agricultural pesticide and the formation of the Environmental Protection Agency in the USA. DDT is still used to control disease-bearing insects such as mosquitoes, but in many parts of the world it is no longer authorized for use on crops.

Carson did not focus solely on DDT but also looked at other pesticides, at how residues built up in foods intended for human consumption, and at how insect disease vectors were rapidly gaining resistance to some of the insecticides that were used with great vigour. She pointed out the carcinogenic nature of many pesticides, and argued for the use of biotic means of pest control – such as introducing natural predators, or making the environment inhospitable to the pests.

Gaia hypothesis

A popular image of the 'living Earth' is the Gaia hypothesis that was proposed by James Lovelock in 1979. The central theory is profoundly ecological. Lovelock

James Lovelock believes humankind faces a serious threat because of the damage people have done to the ecosystem.

sees the Earth as a 'living system' in which all organisms interact not only with one another, but also with the physical attributes of Earth – its geology, climate and atmosphere. The Gaia hypothesis proposes that 'the climate and the composition of the Earth always are close to an optimum for whatever life inhabits it.' The American evolutionary biologist Lynn Margulis, who worked with Lovelock to develop the theory, stressed that Gaia is not an organism, but 'an emergent property of interaction among organisms'.

Lovelock has pointed to such examples as the Great Oxygenation Event, to show the entire Earth-system adapting. About 2.3 billion years ago, cyanobacteria produced sufficient oxygen through photosynthesis to change the chemistry of the atmosphere, so enabling the development and diversification of aerobic life. Opponents point out that although this supports the idea that organisms shape the abiotic environment, it

'The Gaia theory says that the temperature, oxidation state, acidity and certain aspects of the rocks and waters are kept constant, and that this homeostasis is maintained by active feedback processes operated automatically and unconsciously by the biota.'

James Lovelock, 1988

contradicts the Gaian idea that the Earth works to maintain or produce optimum conditions for the prevailing life forms, as the Great Oxygenation Event caused a mass extinction of the microorganisms living at the time.

At the crux of the Gaia hypothesis today is whether the Earth is and can remain self-regulating and self-healing, or whether the rapid changes wrought by human activity tax 'Gaia' to the limit. The idea that biotic and abiotic realms affect each other is no longer in doubt.

Wood-wide web

In the early 21st century, scientists discovered a level of community-wide symbiosis which is quite astonishing and suggests we might have a great deal more still to learn about the intricacies of ecosystems. Suzanne Simard at the University of British Columbia has studied an area of ponderosa pines in British Columbia, just 30m (33yd) square. In it she has mapped the positions and connections of Douglas fir trees and two types of fungus, called mycorrhizae, that live in their roots. The fungi are effectively used as a transport mechanism between the trees, passing chemical 'messages' and even being used to transfer nutrients between them. This is achieved by water and sugar passing into the fungi, some of which the fungi use as food and some of which they transport and deliver to trees in need. Dying

FUNGI EVERYWHERE

The fungi involved in the wood-wide web are everywhere and there are hundreds of varieties of them. They connect the roots of plants through the soil with tendrils. The fruiting bodies of the fungi are mushrooms, toadstools and underground truffles that can be either eaten by animals or cast their spores to the wind.

trees will offload their nutrients for the benefit of others, and established trees share with new trees during periods when they might struggle.

Cooperation goes even further, undermining the principles of Darwinian competition between species. The Douglas fir, an evergreen tree, has a cooperative set-up with paper birch, a deciduous tree. In summer, paper birch supplements fir saplings, providing sugar at a time when they are overshadowed by the tree canopy. In winter, when the birch has lost its leaves, the fir community passes nutrients back to the birch trees. This active nurturing,

WE ARE LEGION

You might feel secure in your identity as a human organism, but you too are an ecosystem. We are all host not only to microscopic parasites that live everywhere from our eyelashes to our guts, but also to billions of microbes that help our bodies to function. Your gut fauna – the microbes in your intestines – help to digest food and keep you healthy, for instance. There are more cells in your body that are not you, but other microbes, than cells that are your own tissues – ten times as many, according to research published in 2007. Indeed, recent research suggests that sometimes the cravings we have for various foods might be prompted by microorganisms that live in us rather than any need we, the hosts, have. We are just as much an ecosystem as any woodland canopy.

even between species, helps to build and maintain a healthy community. In terms of human activity, it means that logging only the largest trees might be more damaging than it appears, as these are providing protection and food for the younger trees.

The role of mycorrhizae in other plant communities is somewhat similar. It can be used to pass alarm signals between plants; a plant that is being grazed or damaged can send a signal either via fungi or by airborne chemicals, or both, which alerts other plants to the danger. Plants receiving the message then increase their defences (toxic or unpleasant-tasting chemicals). Scientists still don't know the motivation: is a tree altruistically sharing carbon? Is it storing it for safe-keeping in its roots? Is it opportunistically taken by the fungi and by other trees? Are

the fungi transferring it to other healthy trees to safeguard their own survival? The matter of agency in organisms such as plants and fungi is difficult to unravel. It is dangerous to use terminology that suggests plants have intentions and a form of psychology, but it is difficult to avoid. The term plant neurobiology, coined in 2005, has attracted criticism as it seems to suggest plants might feel and even think. Plant 'intelligence' is a contentious new idea still to be fully explored.

Forward in time

In the second half of the 20th century and the first years of the 21st century, biology has become a very high-tech science. It has formed close alliances with many other disciplines, making it difficult to say where biology begins and ends. This book has not had space to explore the rapid developments in the applications of genetics, nor to look at the advances in biochemistry that have helped our understanding of what happens within cells.

The plant and animal kingdoms are so vast that we have had the chance to look at biologists' work with just a few organisms among the many millions there are. Biology is an immense subject, of course, for it encompasses all the life-forms on Earth, over the period of 3.5 billion years since life began. Its story is too large and complex to give more than the briefest overview in a short book. But we have had the chance to trace the broad sweep of humankind's study of life. We have seen how perception of the natural world has shifted from one of fixity and hierarchy to recognize instead the shifting complexity of the web of interdependent organisms. We have seen humankind's place reviewed: we no longer consider ourselves rightful masters of a world made solely for our use, but one of myriad organisms that must live together. And we have seen how errors and supernatural explanations have been replaced by accurate knowledge rooted in scientific rigour and diligent enquiry.

But let's not be complacent. There are many mysteries yet to uncover and two of the most crucial questions in biology still puzzle the greatest minds.

Living or not?

We have an innate sense that things which grow, feed, perhaps move, and reproduce are different from things which don't. They have been the subjects of biology. But the line can be surprisingly hard to draw.

We still don't really know what makes something alive. Scientific opinion is divided over whether viruses are living or not. Yet (if we say they are), viruses are the most abundant life, possibly outnumbering all others by a factor of 10. In the oceans, they represent an estimated 94 per cent of

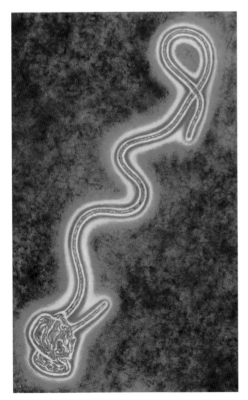

A transmission electron micrograph of a single Ebola virus.

life-forms (but only 5 per cent of biomass because they are so small).

Most people might think that something which can reproduce is alive and something which cannot is not alive. But self-replicating chemicals are again blurring the distinction between living and non-living things. It is now possible to build a virus from scratch, starting with just the chemical components. Biology has yet to locate whatever distinguishes life.

All from nothing?

We have come full circle to considering once again the classification of life with which the Ancient Greeks began. While Aristotle

Only 230,000 marine species are known – there could be anywhere between 1 and 10 million.

divided the natural world into animal, vegetable and mineral, we have tended to keep the mineral element separate. But as archaeobiologists investigate the origins of life on Earth, they find that perhaps it must, after all, have sprung spontaneously from non-living matter, simple chemicals milling around in just the right conditions.

So the most fundamental questions that biology can ask – 'What is life?' and 'Where did it come from?' – remain unanswered after more than 2,500 years of study. There is plenty more of the story yet to be written.

Index

A

Account of the Breeding of Worms in Human Bodies, An (Bois-Regard) 99
Adam, Salimbene di 64
al-Jahiz 187, 188
al-Nafis, Ibn 58–9
Albert of Saxony 139
Alcmaeon of Croton 47–8
Alexander, Annie 195
Altmann, Richard 104
anatomy
    and dissection 47–50
    in Ancient Greece 47–8
    illustrations for 50–3
    mathematics in 55
    mechanistic view of 54–5, 56–8
    in Middle Ages 49–51
    of plants 71–2
    and vivisection 53
Anatomy of Plants, The (Grew) 72
Anaxagoras 63
Anaximander 129
Ancient China
    on evolution 131, 138, 139
    on taxonomy 15
Ancient Greece
    on anatomy 47–8
    on ecology 178–80
    on evolution 129–31, 138
    on physiology 47–8
    on plants 70
    on reproduction 108–9, 112–14, 116
    on taxonomy 15–19, 21, 24
Animal Ecology (Elton) 188
animals
    taxonomy of 15–18, 38–40
Anning, Mary 143
Aquinas, Thomas 25, 114, 132
Aristotle 8, 45
    on anatomy 48
    on ecology 180, 181
    on evolution 130
    influence on Middle Ages 24, 25
    on physiology 48
    on reproduction 108–9, 112–14, 115, 116, 118
    on taxonomy 15–18, 38–9
Arnold, William 78

Art Forms in Nature (Haeckel) 38, 39
Ashmole, Elias 29
Astbury, William 166
Athenaeus 109
Audubon, John James 35
Avery, Oswald 165–6
Avicenna 139

B

Bacon, Francis 182
Bacon, Roger 8, 22
Baer, Karl Ernst von 122–3
Ballestero, Joana and Melchiora 50
barnacle geese 23
Bary, Heinrich de 189
Basil of Caesarea 131
Bassi, Agostino 99, 100
Bateson, William 163, 165
Beadle, George 169
Beaumont, William 66
Becher, Johann 75
Beijerinck, Martinus 102–3
Beneden, Edouard van 159–60
Berthelot, Marcellin 100–1
bestiaries 21–3
Betzig, Eric 105
Bichat, Marie François 96
biogeography 182–7
biological museums 185
Birds of America (Audubon) 35
Bismarck, Otto von 99
blood circulation 58–60, 88
Bois-Regard, Nicolas Andry de 85, 98–9
Bonnet, Charles 74, 119–20, 121
Borelli, Giovanni 56–7, 62
Bosch, Carl 191
Boussingault, Jean-Baptiste 66, 191
Boveri, Theodor 161
Boyle, Robert 61
Bradley, Richard 99
Brookes, Richard 143
Brown, Barnum 155
Brown, Robert 95
Browne, Thomas 183
Buckland, William 142, 144

C

cabinets of curiosities 28–9
Cambrensis, Giraldus 23
Camerano, Lorenzo 188

Camerarius, Rudolf 82–3
Candolle, Augustin de 81
Cardano, Girolamo 8
Carson, Rachel 198–9
Cavalier-Smith, Tom 40
Caventou, Joseph 77
cells 90, 94–8
Chamberland, Charles 102
Chargaff, Erwin 165
Chase, Martha 166
Chatton, Édouard 38
Chillingworth, William 86
Christianismi Restitutio (Servetus) 59
Cicero 131
cladistic model 42
classification *see* taxonomy
Cloquet, Jules Germain 122
Collins, Francis 171
Comparetti, Andrea 77
Conybeare, William 142
Cook, James 196
Cope, Edward Drinker 154–5
Copernicus, Nicolaus 7, 27
Correns, Carl 161
Crick, Francis 147, 166, 167–9
Crookes, Sir William 191
Cuvier, Georges 35–6, 135, 142, 144–5

D

da Vinci, Leonardo 52, 53, 115, 116
Daguerre, Louis 95
Darwin, Charles 9, 142, 144, 145, 158, 192, 194
    biography of 147
    on evolution 146, 148–53
    and genetics 171, 174
    on taxonomy 36–7
Darwin, Erasmus 134
De animalibus (Magnus) 26
Democritus 86
Derbès, August Alphonse 123–4
Descartes, René 54, 62, 117
DeVries, Hugo 161–2
Digby, Sir Kenelm 72
digestion 63–7
Discourse on Earthquakes, A (Hooke) 140
dissection 47–50
Divine Farmer's Herb-Root Classic, The 15

DNA 164–9
Dobzhansky, Theodosius 127, 172–3
Driesch, Hans 125
Dürer, Albrecht 52–3
Dutrochet, Henri 77, 79–80
E
Earl, George Windsor 185
Earthly Venus, The (Maupertuis) 132
ecological niche 194
ecology
    Ancient Greeks on 178–80
    and biogeography 182–7
    and communities of organisms 187–91
    physicotheology 181–2
    and plant distribution 83
    science of 192–203
Edmonstone, John 147
Edwin Smith papyrus 47
eels 109
Ekman, Sven 195
Eldredge, Niles 174
electron microscopes 105
Elton, Charles 188
embryology 113–25
Emerson, Ralph Waldo 138
Emerson, Robert 78
Empedocles 58, 60–1, 129–30, 139
Enquiry into Plants (Theophrastus) 18
Erasistratus 48
Essay on the Principle of Population, An (Malthus) 149
Essays on the Generation of Animals (Harvey) 115
Etymologies (Isidore of Seville) 20, 21
eugenics 169–70
Eukaryotic organisms 38–9
evolution
    Ancient Greeks on 129–31
    and Charles Darwin 134, 146, 148–53
    and creation myths 128–9
    early challenges 132–4
    and fossils 138–45, 153–5
    and genetics 171–5
    intelligent design 130–1
    and Jean-Baptiste Lamarck

134–7, 138
    in Middle Ages 131–2, 139
F
Fabricius, Hieronymus 50, 60
FitzRoy, Robert 146
Flemming, Walther 97–8, 160
fossils 34–5, 138–45, 153–5
Fox, Robert 12
Fracastoro, Girolamo 86, 98
Frank, Albert 189
Franklin, Rosalind 166, 167
Frederick II, Emperor 64
Frosch, Paul 103
G
Gaia hypothesis 199–200
Galapagos Islands 148
Galen
    on anatomy 46, 49, 50, 88
    on blood circulation 58, 60, 61
    on digestion 63
    on reproduction 116
Galton, Francis 169–70
Galvani, Luigi 62–3
genetics
    and cells 159–61
    and DNA 164–9
    and eugenics 169–70
    and evolution 171–5
    genome mapping 171
    and Gregor Mendel 158–9, 161–2
    and Thomas Morgan 162–4
Genetics and the Origin of Species (Dobzhansky) 172–3
genome mapping 171
germs 98–100, 101–3
Gessner, Conrad 27–8, 34–5
Gilbert, Joseph Henry 190
Goldfuss, Georg August 37
Gould, John 148
Gould, Stephen Jay 174
Great Chain of Being 24–6, 33–4
Grew, Nehemiah 71–2, 82–3
Grinnell, Joseph 195
Grosseteste, Robert 8
H
Haber, Fritz 191
Haeckel, Ernst 38, 85, 124, 193
Hales, Stephen 73–4
Haller, Albrecht von 58
Hamburg Hydra 34

Hamm, Stephen 120
Hartsoeker, Nicolaas 120–1
Harvey, William
    on blood circulation 58–60
    on reproduction 115–16, 118–19, 122
Hell, Stefan 105
Helmont, Jan Baptist van 73, 110
Henslow, John Stevens 147
Herophilos 48, 53
Hershey, Alfred 166
Hertwig, Oscar 123, 160
Hesselman, Henrik 194
Hildegard of Bingen 114–15
Hippocrates 46, 98, 112, 179
Histoire naturelle (Leclerc) 35
Historiae animalium (Gessner) 28
History of Animals (Aristotle) 15, 16, 17
History of Plants (Ray) 30
Hitler, Adolf 170
Hobbes, Thomas 54
Hogge, John 38
Hooke, Robert 53, 61–2, 87, 88–90, 139–40
Hooker, Joseph 151
Humboldt, Alexander 183–4
humours 46, 98
Hutton, James 141
Huxley, Thomas 151
I
Idea for a Plant Geography (Humboldt) 184
Ingenhousz, Jan 75–6, 77
insects
    reproduction 119–20
intelligent design 130–1
Isidore of Seville 20, 21
isotopes 78
Ivanovski, Dmitri 102
JK
James I, King 61
Jansen, Hans 86, 87
Jansen, Zacharias 86, 87
Johannsen, Wilhelm 163
Kaman, Martin 78
Kingsbury, Benjamin F. 104
Kircher, Athanasius 29
Knight, Thomas Andrew 81
Knoll, Max 105
Koch, Robert 101, 102
Kölreuter, Joseph 83

Kolthoff, Gustaf 185
Kossel, Albrecht 164
Kosmos (Humboldt) 184
Krakatau 193
L
La Mettrie, Julien Offray de 56
Lamarck, Jean-Baptiste 134–7,
    138
Laue, Max von 166
Lavoisier, Antoine 67, 75, 76
Lawes, John Bennet 190
Leclerc, Georges-Louis 34, 35,
    110, 133–4, 183
Leeuwenhoek, Antonj von 94, 99
    biography of 93
    and microscopes 87, 90–2
    on reproduction 120–1
Leibniz, Gottfried von 26–7
Leidy, Joseph 153
Leopold, Aldo 196–8
Levene, Phoebus 164–5
Lévi-Strauss, Claude 11
Leviathan (Hobbes) 54
Liebig, Justus von 97, 100, 189,
    190–1
Liljefors, Bruno 185
Linnaeus, Carl 70
    on biogeography 182–3
    on ecology 189
    on evolution 133
    on reproduction 121
    on taxonomy 31–3, 34, 37
Lister, Joseph Jackson 94
Lister, Martin 139
Loeffler, Friedrich 103
Lovelock, James 199–200
Lüdersdorff, Friedrich 100
Luther, Martin 7
Lyell, Charles 140, 141–2, 146,
    147, 148, 149, 150
M
Magendie, François 67
Magnus, Albertus 26, 118
Malebranche, Nicolas 117
Malpighi, Marcello 60, 62, 72,
    87–8, 119
Malthus, Thomas 149, 150
Mantell, Gideon 144–5
Margulis, Lynn 199
marine biology 196
Marsh, Othniel 153–5
Maupertuis, Pierre de 132–3

Mayer, Adolf 102
Mayr, Ernst 6, 172, 173–4
McClintock, Barbara 164
Meckel, Johann 125
meiosis 159–161
Mendel, Gregor 158–9, 161–2
Mereschowsky, Konstantin 174–5
Michelet, Jules 151
Micheli, Pier Antonio 111
Micrographia (Hooke) 53, 88–90,
    93, 139
microorganisms
    and Antonj van Leeuwenhoek
    91–2
    and early microscopes 86–7
    early views of 86
    and germs 98–100, 101–3
    in Great Chain of Being 34
    taxonomy of 37–8, 40–1
    viruses 102–3, 202–3
*Microscopic Investigations on the
    Accordance in the Structure and
    Growth of Plants and Animals*
    (Schwann) 96
Middle Ages
    on anatomy 49–51
    on evolution 131–2, 139
    on reproduction 109, 114,
    118
    on taxonomy 20–6
Miescher, Friedrich 164
Miller, Philip 83
mitochondrion 103–4
mitosis 159–61
Moerner, William 105
Mohl, Hugo von 77
Morgan, Thomas 162–4, 172
Muller, Hermann 172, 173, 175
muscles 56–8, 62–3
NO
Nägeli, Karl von 159
Natural History (Pliny the Elder)
    19, 20
Nicolas of Cusa 24
Nirenberg, Marshall 169, 170
ocelloids 104
Oken, Lorenz 128
On the Causes of Plants
    (Theophrastus) 18
On the Fabric of the Human Body
    (Vesalius) 50–1
On the Generation of Animals

(Aristotle) 108
On the Generation of Animals
    (Hippocrates) 112
On the Motion of the Heart
    (Harvey) 59
On the Nature of Things (Isidore
    of Seville) 20
*On the Origin of Species by Natural
    Selection* (Darwin) 36, 124, 147,
    149, 150–1, 153, 158
On the Parts of Animals (Aristotle)
    15, 16, 45
osmosis 79–80
Ostwald, Wilhelm 191
Owen, Richard 38, 39–40, 144, 145
P
Paley, William 131
Pander, Heinz Christian 122,
    123
Paracelsus 115
Park Grass experiment 190, 191
parthenogenesis 119–20
Pasteur, Louis 97, 99–102,
    111–12
Pelletier, Pierre 77
Pepys, Samuel 89
Petri, Julius 101
photosynthesis 74–80
Physiologus 21
physicotheology 181–2
physiology
    in Ancient Greece 47–8
    blood circulation 58–60, 88
    digestion 63–7
    and dissection 47–50
    and humours 46
    mathematics in 55
    mechanistic view of 54–5,
    56–8
    muscles 56–8, 62–3
    plants 72–3
    respiration 60–2, 88
Pigafetta, Antonio 182
plants
    anatomy of 71–2
    Ancient Greeks on 70
    and ecology 83
    nutrition for 73–4, 189–91
    photosynthesis 74–80
    physiology of 72–3
    reproduction 82–3
    taxonomy of 18–19, 31–2,

38–40, 70–1
tropism 80–2
Plato 130, 179
Pliny the Elder 6, 19, 20, 21, 109
Plot, Robert 142, 143
Priestley, Joseph 74–5, 117, 118
Principles of Geology (Lyell) 141
Prokaryotic organisms 38, 39
protists 37, 38
protozoa 37, 38, 43
Pruvost, Melanie 13
Pseudo-Dionysius the Aeropagite 24
Purkinje, Jan 95
R
Ray, John 30, 140
Redi, Francesco 109–10
Religion of Protestants (Chillingworth) 86
Remak, Robert 97
reproduction
    Ancient Greeks on 108–9, 112–14, 116
    embryology 113–25
    fertilization 118–23
    insects 119–20
    Middle Ages on 109, 114, 118
    parthenogenesis 119–20
    of plants 82–3
    and spontaneous generation 108–12
respiration 60–2, 88
Röntgen, William 166
Roux, Wilhelm 125
Royal Society 8, 89, 92, 93
Ruben, Samuel 78
Rudolf II, Emperor 29
Ruska, Ernst 105
S
Sachs, Julius von 77
St Martin, Alexis 66
Sand County Almanac, A (Leopold) 197
Sanger, Fred 171
Santorio, Santorio 55, 73
Saussure, Nicolas de 77, 190
Schleiden, Matthias 77–8, 95–6
Schwann, Theodor 95–7
Sclater, Philip 184–5
scientific method 8

sea urchins 125
Sedgwick, Adam 147
Senebier, Jean 76
Seneca 86, 117
Sernander, Rutger 194
Servetus, Michael 59
Shakespeare, William 107, 109
Shen Kuo 139, 181
Shennong 15
Silent Spring (Carson) 198
Simard, Suzanne 200
Slyvius, Jacobus 50
Smith Bowen, Elenore 68
Sophocles 179
Spallanzani, Lazzaro 65–6, 111, 121
spontaneous generation 108–12
Sprengel, Carl 191
Sprengel, Christian 83
Stanier, Roger 38
Steno, Nicolaus 57, 62, 140–1
Stirn, Georg Christoph 29
Strasburger, Eduard 98, 161
Sturtevant, Alfred 162–3
Stutchbury, Samuel 145
Suess, Eduard 195
Sutton, Walter 162
Swammerdam, Jan 57, 92–4, 119
Swift, Jonathan 90
Systema Natura (Linnaeus) 32
Systematics and the Origin of Species (Mayr) 173, 174
T
Tatum, Edward 169
taxonomy
    in Ancient Greece 15–19, 21, 24
    of animals 15–18, 38–40
    and Carl Linnaeus 31–3, 34, 37
    and Charles Darwin 36–7
    cladistic model 42
    contemporary 42–3
    in 18th century 35–6
    and Great Chain of Being 24–6, 33–4
    and John Ray 30
    of microorganisms 37–8, 40–1
    in Middle Ages 20–6
    of plants 18–19, 31–2, 38–40, 70–1
    prehistoric 13–14

tree of life model 40, 41–2
Thales 129
Theophrastus 18–19, 70, 177, 180, 189
Theory of Earth (Hutton) 141
Thomson, Charles Wyville 196
Topographica Hiberniae (Cambrensis) 23
Tradescant the Elder, John 29
tree of life model 40, 41–2
tropism 80–2
Tschermak, Erich von 161
VW
van Niel, C. B. 38
Vegetable Staticks (Hales) 74
Vernadsky, Vladimir 195–6
Vesalius, Andreas 27, 50–1
Vespucci, Amerigo 27
Virchow, Rudolf 97, 98, 99
viruses 102–3, 202–3
vivisection 53
Volta, Alessandro 62–3
Waldeyer-Hartz, Wilhelm von 98, 160
Wallace, Alfred Russell 150, 185–6
Waller, Robert 90
Wallin, Ivan 175
Wang Mang 49
Warming, Eugen 192, 194
Watson, James 166, 167–9, 170–1
Wedgwood, Emma 147
Wegener, Alfred 186–7
Weismann, August 137–8, 157, 160–1, 171
Whittaker, Robert 38
Willughby, Francis 30
Woese, Carl 40–1
Wolff, Caspar 121–2
Wolgemut, Michael 52
Worm, Ole 29
XYZ
Xenophanes 129
yeast 100–1
Zhuangzi 131
Zoogeography of the Seas (Ekman) 195
zoology
    origins of 27–9
Zoönomia (Darwin) 134

Picture credits